POSTGRES EVERYWHERE

Install on any OS, container, cloud & serverless

Jose Berardo

PostgreSQL Mastery

Postgres Everywhere

Master Postgres Database Setup Across All Major Platforms

Jose Berardo

September 2025

Contents

PostgreSQL Installation Overview: Your Complete Setup Strategy

Welcome to Postgres Everywhere, your comprehensive guide to installing and configuring PostgreSQL across every major platform and deployment method. Whether you're a developer setting up your first development environment, a system administrator preparing production systems, or a cloud architect evaluating managed services, this book provides the practical knowledge you need to get PostgreSQL running anywhere.

Think about it this way: you wouldn't start building a house without first preparing the foundation and gathering your tools. Similarly, a well-planned PostgreSQL installation serves as the bedrock for everything you'll accomplish with this powerful database system. This chapter establishes the strategic framework for making informed installation decisions.

Why Multiple Installation Approaches Matter

You might wonder: "Why not just show me one way to install PostgreSQL and move on?" This is a great question, and the answer reflects the reality of modern database professionals and the diverse environments you'll encounter throughout your career.

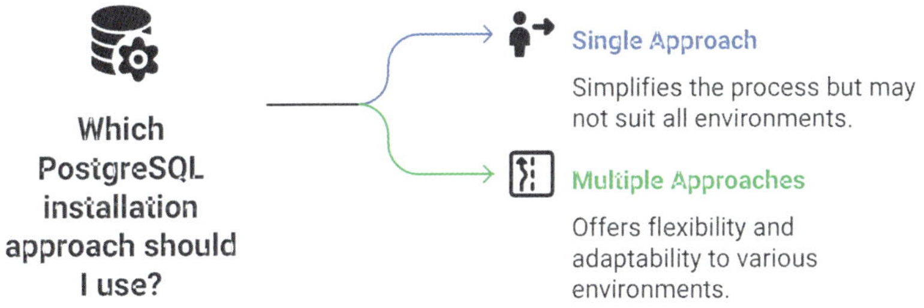

Benefits of learning multiple installation approaches

The Multi-Platform Reality

In today's technology landscape, you'll rarely work in a single-platform environment. Consider these common scenarios:

Development Teams: You might develop on macOS, your colleagues use Windows, and your production servers run Linux. Understanding installation across platforms ensures you can collaborate effectively and troubleshoot issues regardless of the environment.

Career Flexibility: Database professionals who understand multiple platforms are more valuable. Whether you're consulting, changing companies, or your organization migrates platforms, this knowledge travels with you.

Cloud-Native Mindset: Modern applications increasingly span multiple cloud providers and hybrid environments. AWS RDS, Azure Database, Google Cloud SQL, and serverless options like Neon each have their place in the ecosystem.

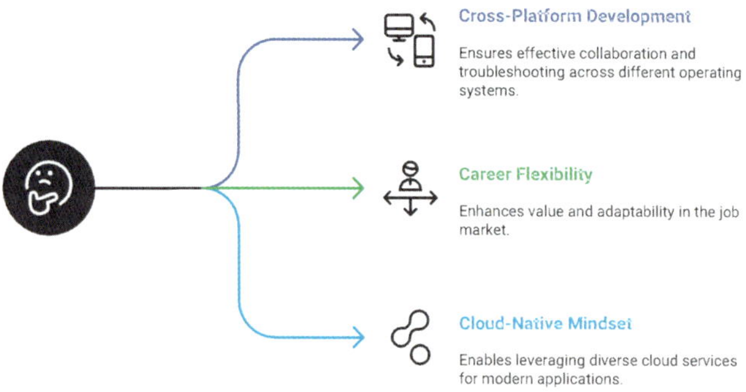

The Multi-Platform Reality

Installation Method Diversity: Strategic Advantages

Our restructured approach covers installation methods strategically rather than repetitively:

Visual Installation (Chapter 2): Perfect for development environments and learning. The EnterpriseDB installer provides identical experiences across Windows and macOS, making it an excellent starting point for newcomers while offering the comprehensive toolset needed for professional development.

Terminal Installation (Chapter 3): Essential for production environments, automation, and DevOps workflows. Package managers like `apt`, `brew`, and `choco` represent modern best practices, and understanding them helps you with far more than just PostgreSQL installation.

Cloud Services (Chapter 4): Managed services eliminate infrastructure overhead and provide enterprise-grade features out of the box. Understanding when to choose RDS over self-managed installations can save organizations thousands of hours and significant costs.

Serverless Options (Chapter 5): Services like Neon, Supabase, and TigerData represent the cutting edge of database-as-a-service offerings, perfect for modern applications that need to scale from zero to millions of users.

Container Fundamentals (Chapter 6): Docker and Kubernetes are now essential skills for database professionals. Whether you're setting up development environments, creating reproducible deployments, or managing microservices, containerized PostgreSQL is everywhere.

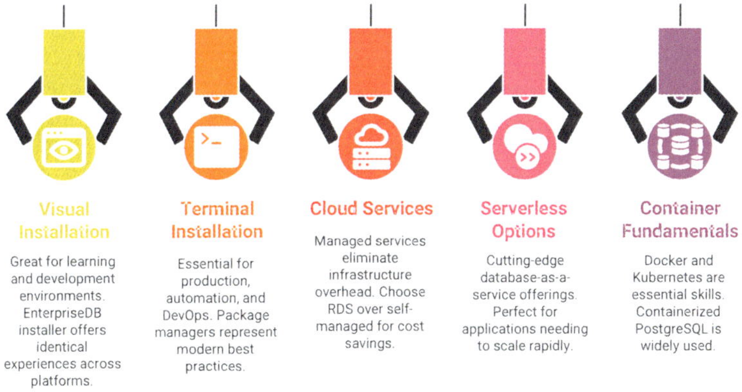

A plethora of installation methods

When to Choose Visual Installation

Before we dive into the actual installation process, let's understand when the visual installation approach makes the most sense for your specific situation.

Choose Visual Installation When: - You're new to PostgreSQL and want a guided, comprehensive setup experience - You're learning PostgreSQL and want all the tools available immediately - You're setting up a development environment on Windows or macOS - You're more comfortable with graphical interfaces than command-line operations - You want the complete PostgreSQL ecosystem in one installation (server, pgAdmin, command-line tools, and some extra tools installed by a feature called Stack Builder)

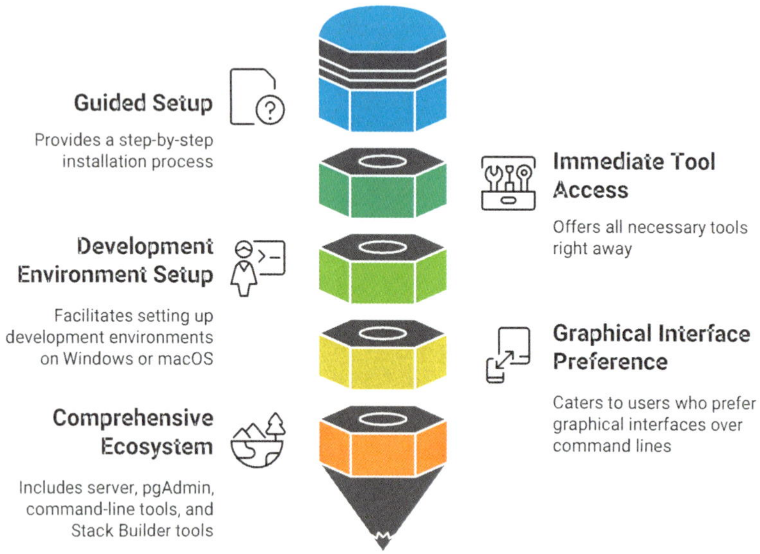

When to choose visual installation

Skip Visual Installation When: - You're setting up production servers (we'll cover this in our next chapter) - You're working in headless/server environments without GUI - You need specific customization or minimal installations - You're automating deployments or using infrastructure as code

When to avoid visual installation

Think of visual installation as your PostgreSQL starter pack - it gives you everything you need to begin your PostgreSQL journey with minimal technical friction.

Why Terminal Installation Matters

Let's understand why mastering terminal-based installation is crucial for your PostgreSQL journey:

Developer Workflow Integration: Terminal installations integrate seamlessly with automation scripts, deployment pipelines, and configuration management tools that you'll encounter in professional environments.

Reproducible Environments: Commands can be scripted, shared, and version-controlled, ensuring consistent installations across teams and environments.

Better Updates and Maintenance: Package managers excel at dependency resolution, security updates, and clean uninstallation - crucial for production environments.

Learning Investment: Understanding package managers helps with far more than PostgreSQL. These skills transfer to installing development tools, libraries, and system components across your entire tech stack.

Version Selection: Looking Ahead to PostgreSQL 18

PostgreSQL follows a predictable annual release cycle that makes version selection straightforward once you understand the patterns. As we record this course in 2025, we're in an exciting period with PostgreSQL 17 mature and stable, and PostgreSQL 18 approaching release with groundbreaking features.

PostgreSQL 18: A Game-Changing Release

PostgreSQL 18, currently in beta, represents one of the most significant advances in PostgreSQL's recent history. Here's why it matters for your learning journey:

Revolutionary Performance Improvements: - **Asynchronous I/O**: The new AIO subsystem delivers 2-3x performance improvements for read-heavy workloads - **Better Function Optimization**: SQL and PL/pgSQL functions now benefit from smarter execution planning - **Enhanced B-tree Operations**: Skip scan support improves query performance when not all index columns are specified

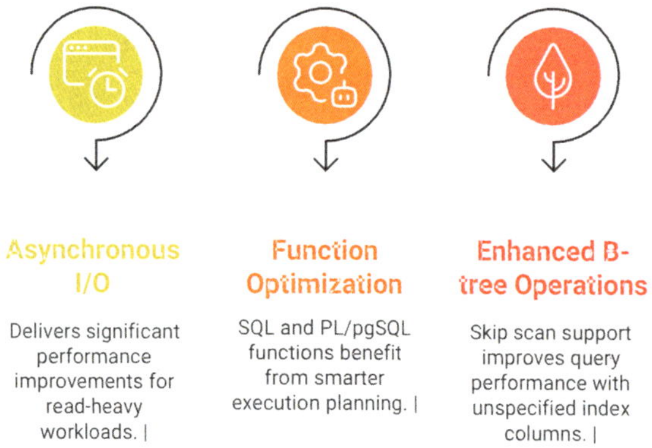

Postgres 18 Performance Improvements

Enterprise-Grade Features: - **OAuth 2.0 Authentication**: Finally bringing PostgreSQL in line with modern enterprise identity standards - **Enhanced Logical Replication**: DDL commands now replicate automatically, making multi-region deployments much more practical - **Virtual Generated Columns**: Compute-on-read columns that save storage space and write overhead

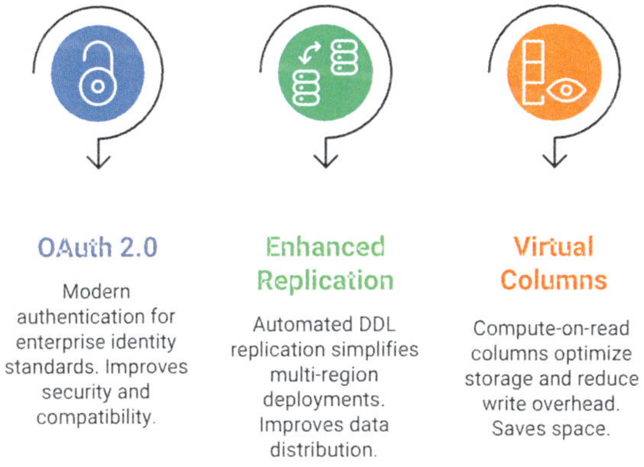

Postgres 18 Enterprise-Grade Features

Developer Experience Enhancements: - **Enhanced MERGE Statements**: More SQL-standard compliant UPSERT operations - **JSON_TABLE Support**: Better handling of semi-structured data - **Improved RETURNING Clauses**: More flexibility in getting old and new values from DML operations

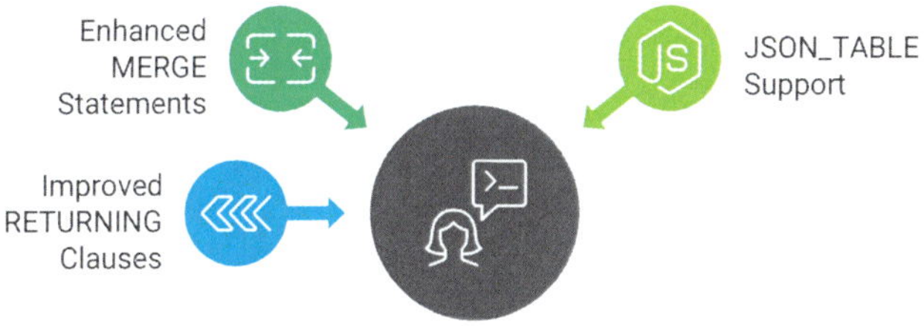

Postgres 18 Enhancing Developer Experience

Version Selection Strategy

For your PostgreSQL installation and learning journey, here's the recommendation:

For Learning (Now): PostgreSQL 17 is your best choice. It's stable, well-documented, and comes with the full set of essential server components and tools you need for setup and initial exploration.

For New Projects: PostgreSQL 17 for immediate production needs, with a roadmap to evaluate PostgreSQL 18 once it reaches general availability.

For Existing Projects: Follow your organization's upgrade policies, but start planning for the performance benefits of PostgreSQL 18's asynchronous I/O.

System Requirements: Beyond the Minimum

When planning to install and run PostgreSQL 17 effectively, it's important to account for realistic operational needs rather than just minimums.

Recommended Hardware

- **Processor:** At least a 1 GHz CPU is required for basic operation, but 2 or more cores are recommended for concurrent workloads to allow smooth handling of simultaneous queries and background maintenance processes. PostgreSQL runs efficiently on modern x86_64 processors and ARM64 architectures.
- **Memory:** A minimum of 2 GB RAM is the baseline to allow PostgreSQL and the operating system to operate reliably. However, 4 GB or more is strongly recommended for most environments. This allows sufficient memory for caching data, maintaining multiple connections, and managing complex queries without excessive swapping or performance degradation.
- **Storage:** While PostgreSQL can be installed on systems with as little as 512 MB of disk space, at least 10 GB of free disk space on a fast SSD or NVMe storage device is recommended to accommodate the database cluster files, transaction logs, and temporary files. Adequate disk throughput and low latency are critical for database performance.

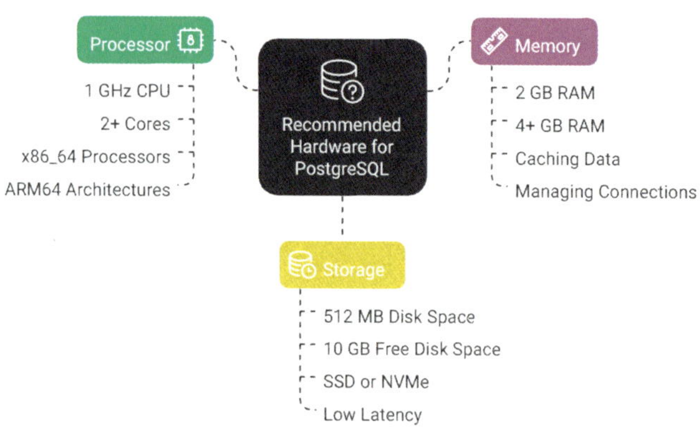

Postgres minimum recommended hardware configuration

Operating System and Software

PostgreSQL 17 is widely supported on major operating systems such as Linux (various distributions), Windows 10 and Server 2019 or later, and macOS (version 10.15 or later). File system choice impacts performance: ext4 or XFS are preferred for Linux, APFS for macOS, and NTFS on Windows.

Dependencies include a C compiler compliant with C99, GNU Make 3.81 or later (or equivalent), and essential libraries such as readline, zlib, and others depending on installed features.

Additional Considerations

- For development and light workloads, virtual machines or cloud instances with 2 CPU cores, 2 GB RAM, and SSD storage are sufficient to run PostgreSQL reliably.
- Production environments should scale resources based on workload characteristics, with monitoring to identify bottlenecks in CPU, memory, storage I/O, and network.
- Using SSDs or NVMe for storage significantly improves transactional throughput and latency compared to spinning disks.
- Configurations such as shared_buffers should be tuned to utilize available memory effectively.

For detailed and up-to-date system requirements, consult the official PostgreSQL documentation: https://www.postgresql.org/docs/17/install-requirements.html and vendor-specific resources.

Our Strategic Learning Path

This book is designed as a choose-your-own-adventure approach that respects your time and specific needs:

Chapter 2: Visual Installation & Tools (Windows and macOS)

Perfect if you're: - New to PostgreSQL and want a guided experience - Setting up a development environment on Windows or macOS - Interested in GUI tools like pgAdmin for database management

The EnterpriseDB installer provides virtually identical experiences across platforms, making this an efficient way to understand the complete PostgreSQL ecosystem.

Chapter 3: Terminal Installation & Tools (Linux, macOS, Windows)

Essential if you're: - Planning to work in production environments - Interested in automation and DevOps practices - Want to understand package managers and system-level configuration

This chapter covers the universal principles that apply regardless of your specific platform.

Chapter 4: Cloud PostgreSQL Options (AWS, Azure, GCP)

Crucial for understanding: - Managed database services and their trade-offs - Enterprise-grade features like automated backups and high availability - Cost optimization and scaling strategies

Chapter 5: Serverless Options (Neon, Supabase, TigerData)

Explores the cutting edge: - Serverless database architecture and its benefits - Modern application development patterns - Specialized use cases like time-series data and real-time applications

Chapter 6: Container Fundamentals (Docker/Kubernetes)

Covers modern deployment: - Containerized database development and testing - Kubernetes orchestration for PostgreSQL - DevOps integration and CI/CD pipelines

Chapter 7: PostgreSQL Tools & Interfaces

Unifies everything: - Deep dive into `psql` and its advanced features - Comprehensive pgAdmin 4 walkthrough - Comparison of popular GUI tools and their use cases - Introduction to other tools like DBeaver and Visual Studio Code

Installation Support and Troubleshooting Philosophy

Let's address the elephant in the room: installation issues happen. The diversity of system configurations, versions, and existing software means you might encounter unique challenges. Here's how to approach troubleshooting systematically:

Before You Start

1. **Document your environment**: Operating system, version, existing software, and system specifications
2. **Choose your path**: Decide whether you want the visual experience or command-line approach based on your comfort level and goals
3. **Plan your time**: Allow 30-60 minutes for installation and basic configuration

When Issues Arise

1. **Check PostgreSQL community forums first**: Many installation challenges have been documented and solved by the community
2. **Consult official documentation**: PostgreSQL's documentation is comprehensive and well-maintained
3. **Search systematically**: Use specific error messages in your searches rather than general terms
4. **Document solutions**: Share your solutions in the forums to help others

Version Flexibility

PostgreSQL's commitment to backward compatibility means the core concepts remain consistent across versions. If you're using a newer version than demonstrated:

- Installation steps remain largely the same
- SQL syntax is backward compatible
- New features are additions, not replacements
- Configuration concepts remain consistent

Looking Ahead: Your PostgreSQL Journey

This foundational chapter prepares you for practical PostgreSQL implementation across diverse environments. The installation methods covered in this book enable you to:

Deploy PostgreSQL Anywhere: From local development environments to enterprise cloud deployments, you'll have the knowledge to choose and implement the right installation approach for any situation.

Optimize for Your Use Case: Understanding the trade-offs between visual installers, package managers, cloud services, and containerized deployments helps you make informed decisions that align with your specific requirements.

Build Professional Skills: The installation techniques you master here form the foundation for advanced PostgreSQL administration, performance optimization, and production deployment strategies.

Adapt to Changing Requirements: As your PostgreSQL needs evolve from development to production to enterprise-scale deployments, you'll have the flexibility to migrate between installation methods while maintaining your existing knowledge and skills.

Key Takeaways

As we conclude this foundational chapter, remember these crucial principles:

1. **Choose Based on Goals:** Visual installation for learning and development, terminal installation for production and automation, cloud services for enterprise-grade features
2. **Embrace Multiple Platforms:** Understanding diverse installation methods makes you adaptable to any environment and valuable as a professional
3. **Plan for Evolution:** Your PostgreSQL installation needs will grow and change—the methods in this book provide the flexibility to adapt
4. **Think Strategically:** Installation is just the beginning—choose methods that align with your long-term PostgreSQL journey
5. **Document Everything:** Good installation practices and troubleshooting skills transfer to countless other technologies

The following chapters will guide you through each installation method with practical, step-by-step instructions. You'll gain hands-on experience with visual installers, command-line tools, cloud platforms, serverless architectures, and containerized deployments. By the book's end, you'll have the expertise to deploy PostgreSQL confidently in any environment.

This foundation in PostgreSQL installation represents more than technical knowledge—it's your gateway to leveraging one of the world's most powerful database systems. Whether you're building your first application or architecting enterprise-scale solutions, the installation mastery you develop here will serve you throughout your PostgreSQL journey. The next chapter begins our hands-on exploration with visual installation tools, where theory transforms into practical expertise.

Additional Resources

- **PostgreSQL Mastery**. Editor's education platform
- **PostgreSQL 18 Beta Information**: Official Release Notes
- **PostgreSQL Architecture Deep Dive**: Understanding PostgreSQL Internals
- **Cloud Platform Decision Guide**: Choosing AWS Database Services
- **PostgreSQL Community**: Installation Guides Wiki

Visual Installation & Tools (Windows and macOS)

Having explored the overview and strategic considerations for PostgreSQL installation in Chapter 1, we now turn to the practical implementation of visual installation methods. This chapter focuses on the EnterpriseDB installer, which provides the most user-friendly and comprehensive way to get PostgreSQL running on both Windows and macOS systems. Building upon the foundation established in the previous chapter, we'll walk through each step of the visual installation process.

This chapter also brings up a bonus installation alternative for macOS users, the Postgres.app macOS native application that runs the server while the app is open.

Downloading the EnterpriseDB Installer

The first step in your PostgreSQL installation journey is obtaining the official installer from the PostgreSQL website. Both Windows and macOS follow the same initial process, with only minor platform-specific differences.

Accessing the Download Page

Navigate to postgresql.org and locate the download section. You can either click the prominent "Download" button on the homepage or select "Download" from the main navigation menu. This will take you to the main download page at postgresql.org/download where you'll see options for different operating systems.

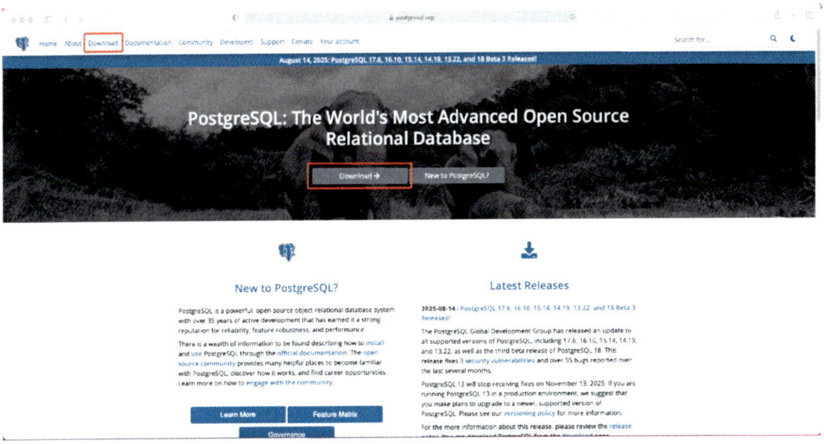

PostgreSQL homepage showing download options

Platform Selection

The Download page offers several alternatives:

- **Linux**: Opens a sublist of Linux distributions: Debian, Red Hat/Rocky/AlmaLinux, SUSE, Ubuntu, or Others. They all brings you to a page for more instructions on how to install via package management. Which we'll cover on next chapter
- **macOS users**: Brings you to the macOS specific download page
- **Windows**: Brings you to the Windows specific download page
- **BSD and Solaris**: Function more similar to Linux

As a popular open-source software, you can always alternatively click on the Sources link on Quick Links to download the source code of any version. From v1.08 to v18beta3 (as of now).

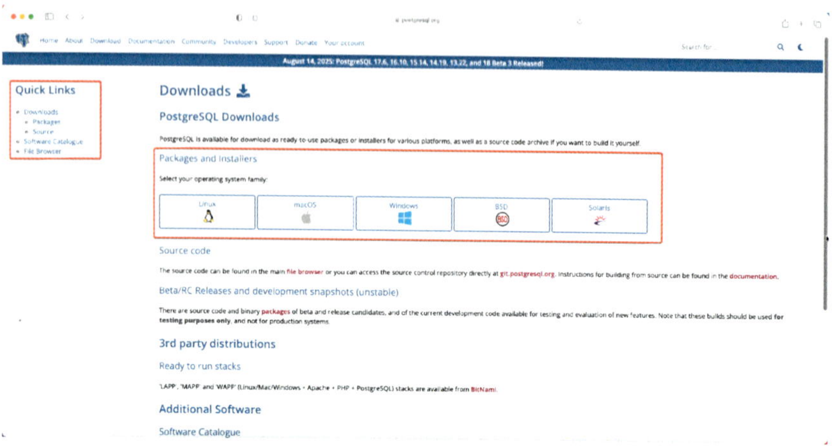

Download page

To select a visual installation, click on either "Windows" or "macOS" depending on your system. Both will display an intermediate page on either /download/windows or /download/macos.

- The Windows page: postgresql.org/download/windows, displays all supported versions.
- While the macOS page: postgresql.org/download/macosx, not only displays pretty much the same versions but also some alternative installation methods:
  - The Postgres.app a no brainer way to simply run Postgres directly from the menubar, without even need to install
  - The Homebrew method, which we cover in the next chapter, as one of the terminal installation methods
  - MacPorts and Fink: Two open source initiatives for compiling and installing things on macOS or other Unix like systems

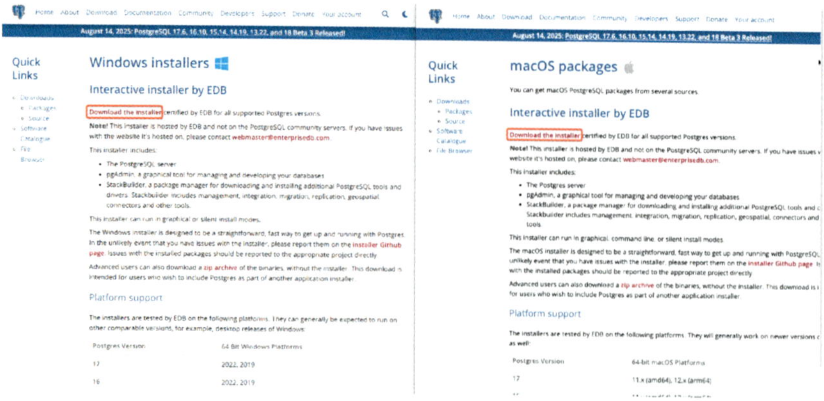

Side-by-side view of Windows and macOS download pages

Accessing the EnterpriseDB Installer

On both platforms, click "Download the installer" link. This will redirect you to the EnterpriseDB.com download page, regardless of which platform you selected. This unified page serves as the official distribution point for the visual EnterpriseDB installer.

The download page presents multiple columns for different operating systems. While Linux options redirect back to PostgreSQL.org, the Windows and macOS columns provide direct access to the visual installer downloads.

The download page will show options for: - PostgreSQL Server versions (17.x, 16.x, 15.x, etc.) - Platform-specific installers (Windows 64-bit, macOS)

Version Selection and Download

At the moment I'm writing this book, PostgreSQL 17.6 is recommended as it offers the best balance of stability and modern features. However, you can also choose PostgreSQL 16.10 for longer community support history, or even try PostgreSQL 18 when available (though it may still be in beta).

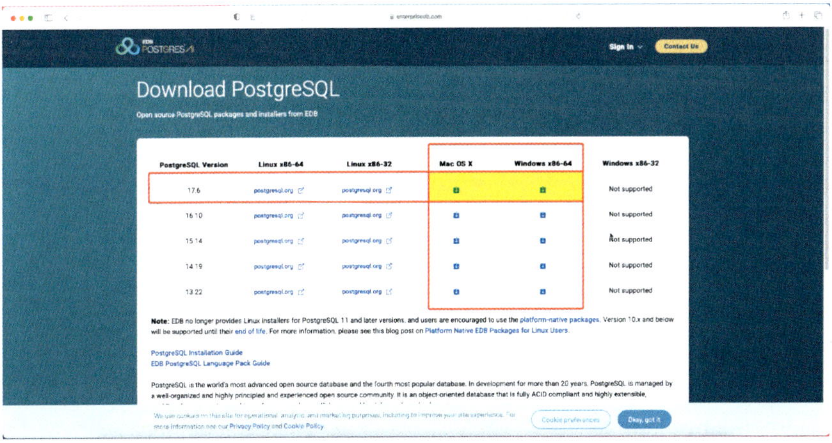

EnterpriseDB download page showing version matrix

Select your platform column and click on your preferred version. The installer files are typically:

- **Windows**: postgresql-17.x-windows-x64.exe (approximately 200-400MB)
- **macOS**: postgresql-17.x-osx.dmg (similar size)

Save the installer to your preferred download location. Ensure you have administrative privileges on your system, as you'll need them for the installation process.

Installation Process Overview

The installation wizard provides nearly identical experiences on both Windows and macOS platforms. However, there are important differences in how each operating system handles security permissions and installer execution that can significantly impact your installation experience.

Understanding Administrative Privileges

Before launching the installer, it's important to understand why administrative privileges matter for database software installation:

What Administrative Privileges Provide: - Permission to write to protected system directories like `C:\Program Files` (Windows) or `/Library` (macOS) - Ability to create and configure system services that start automatically - Access to modify system registry entries (Windows) or system configuration files (macOS) - Rights to create dedicated user accounts for the PostgreSQL service - Permission to configure network ports and firewall settings

Why PostgreSQL Needs These Privileges: - PostgreSQL installs as a system service that runs continuously in the background - The database server needs to write to protected system locations for binaries and configuration - Network services require elevated permissions to bind to ports (like the default `5432`) - Proper security isolation requires creating dedicated service accounts

Launching the Installer on Windows

On Windows, you have two main approaches to launching the installer:

Method 1: Standard Double-Click Simply double-click the downloaded .exe file (typically named `postgresql-17.x-windows-x64.exe`). Windows will display security warnings since this isn't a Microsoft Store application. Click "Install anyway" when prompted, then "Yes" when the User Account Control (UAC) dialog appears asking "Do you want to allow this app to make changes to your device?"

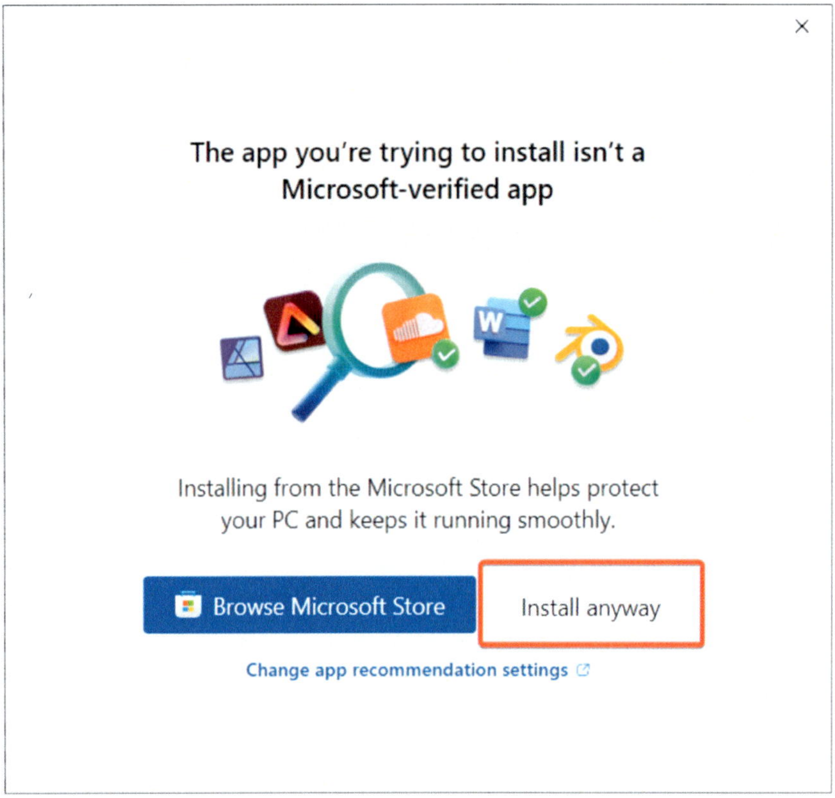

Microsoft Store warning screen

Method 2: Explicit Administrator Rights (Recommended) Right-click the downloaded .exe file and select "Run as administrator." This approach is recommended because: - It ensures the installer runs with full elevated privileges from the start - Prevents potential permission-related issues during installation - Guarantees proper system service installation and configuration - Avoids complications with files that might be created with mixed permission levels

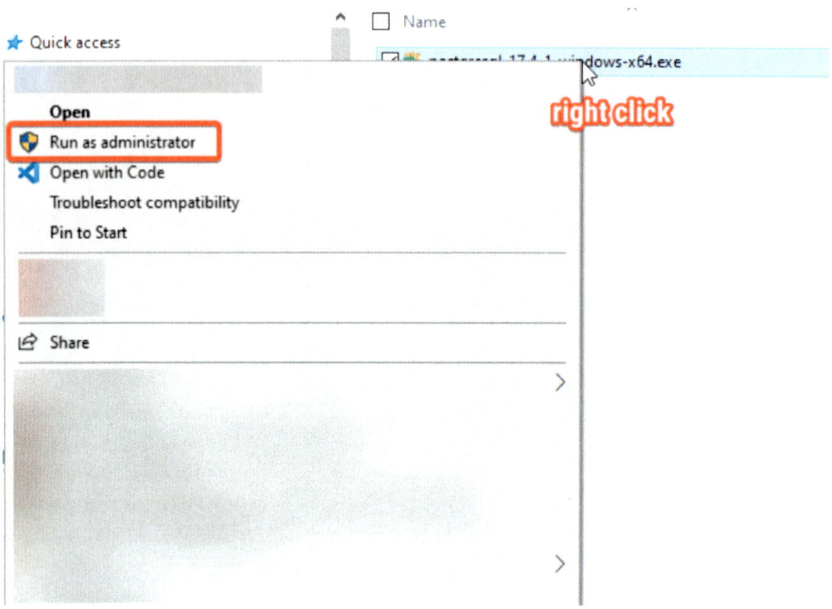

Run as administrator option

When might you need "Run as administrator": - If you encounter "Access denied" errors during installation - When installing on corporate or managed Windows systems with strict security policies - If the standard launch fails to install system services properly - When you need to install to non-standard locations that require elevated access

Launching the Installer on macOS

macOS handles administrative privileges through its built-in security mechanisms:

1. Double-click the downloaded .dmg file to mount it (typically named `postgresql-17.x-osx.dmg`)
2. Double-click the PostgreSQL installer icon inside the mounted disk image
3. If macOS displays security warnings about the app being downloaded from the internet, it most likely say that Apple checked it for malicious software and none was detected. So you are good to click **Open**.
4. macOS will prompt for your administrator password - enter it when requested

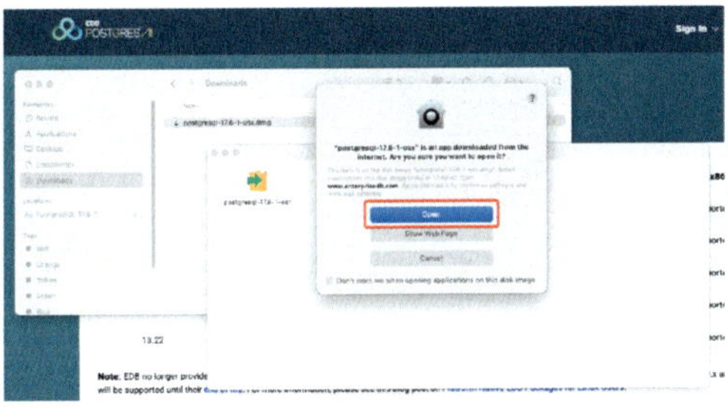

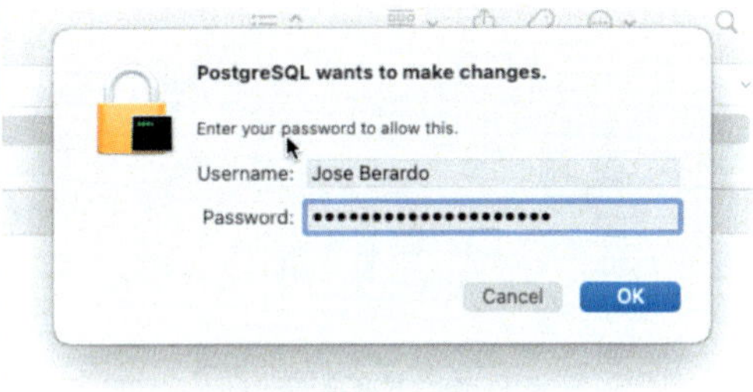

In case it doesn't allow you to proceed, try these steps:

1. Go to System Preferences > Security & Privacy
2. Click "Open Anyway" next to the PostgreSQL installer message
3. Confirm your choice in the subsequent dialog

Why These Security Steps Matter

Both Windows and macOS implement these security measures to protect users from potentially malicious software. However, PostgreSQL is legitimate database software that requires system-level access to function properly. Understanding these security prompts helps you:

- Recognize when elevated privileges are necessary vs. optional
- Troubleshoot installation issues related to insufficient permissions
- Ensure your PostgreSQL installation has all required system access
- Maintain security best practices while installing essential software

Installation Wizard Walkthrough

Both platforms present identical installation wizard screens once security permissions are granted. This parallel approach (on Windows and macOS) demonstrates the consistency of the EnterpriseDB installer experience.

Welcome and Initial

The installation wizard opens with a welcome screen confirming you're installing the official PostgreSQL distribution from EnterpriseDB. Click "Next" to proceed.

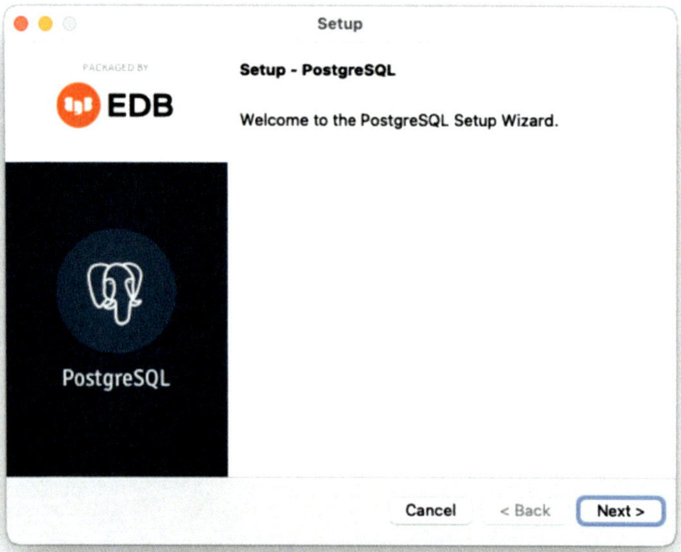

Welcome screen on macOS

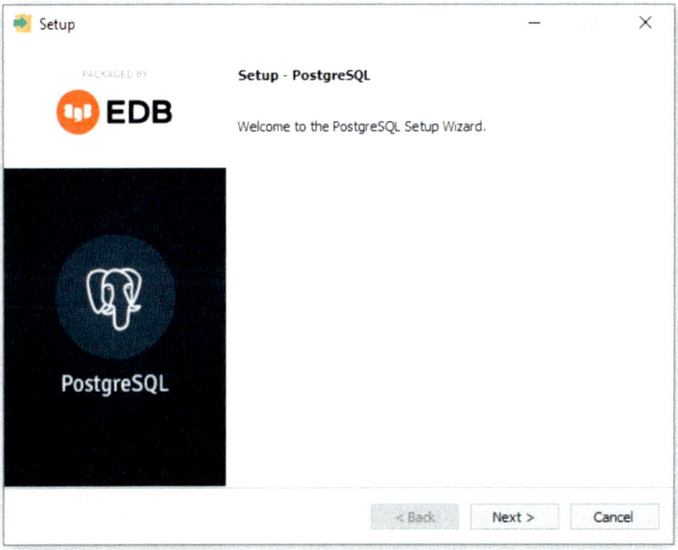

Welcome screen on Windows

Directory Selection:

The installer will show different screens depending on whether this is a fresh installation or an upgrade:

- **macOS default**: `/Library/PostgreSQL/17`
- **Windows default**: `C:\Program Files\PostgreSQL\17`

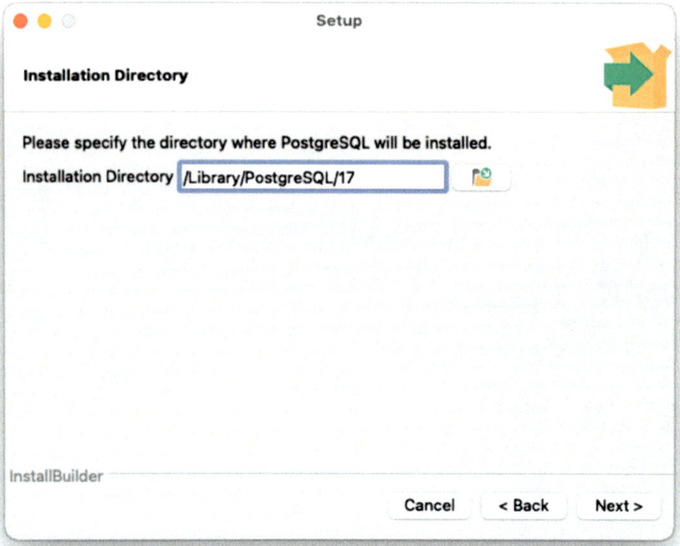

Example of fresh installation screen on macOS

Unless you have specific requirements, accept the default location. If you need to change it: - Choose a path without spaces for compatibility - Ensure the drive has at least 1GB of free space - Avoid system-protected directories

Click **Next** to proceed.

Upgrade Installation: If you already have PostgreSQL installed, the installer will detect this won't display the directory selection screen. Instead, it'll jump straight to the next screen: Select Components.

Component Selection

This is one of the most important screens in the installation process. You'll decide which components to install, and understanding these choices will serve you well throughout your PostgreSQL work.

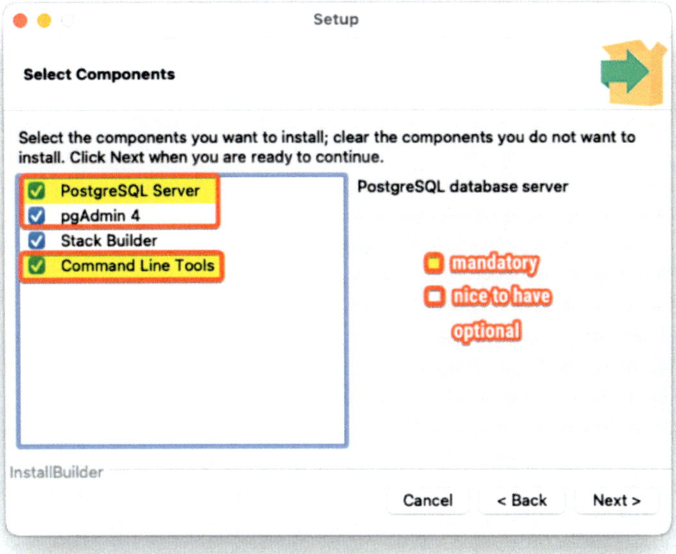

Select Components screen of a fresh installation shown on macOS

The installer presents four main components:

PostgreSQL Server (Required) This is the core database engine — the heart of PostgreSQL. This component includes: - The PostgreSQL database server process - Core SQL engine and query processor - Built-in authentication and security systems - Standard PostgreSQL extensions

pgAdmin 4 (Highly Recommended) PostgreSQL's premier graphical administration tool, serving as the control panel for your databases. pgAdmin 4 provides: - Visual database management interface - Query editor with syntax highlighting - Performance monitoring and statistics - User and permission management tools - Backup and restore functionality - Database schema visualization

Stack Builder (Optional) Your gateway to the PostgreSQL ecosystem—essentially an app store for PostgreSQL extensions and tools. Stack Builder enables installation of: - Additional PostgreSQL extensions (PostGIS for geographic data, pg_stat_statements for query analysis) - Database connectivity drivers (ODBC, JDBC) - Additional administrative tools - Language bindings and interfaces

Command Line Tools (Essential) The command-line utilities that power PostgreSQL operations, including: - `psql`: The interactive PostgreSQL command-line client - `pg_dump` and `pg_restore`: Backup and restore utilities - `createdb` and `dropdb`: Database creation and deletion tools - `pg_ctl`: PostgreSQL server control utility

Recommendation: Select all components for a complete installation. This gives you maximum flexibility and learning opportunities. The additional disk space (approximately 200-300MB total) is minimal compared to the functionality you gain.

Click **Next** to proceed.

Select Components on Upgrade Installation When your installation is automatically detected as an upgrade (there's a previous installation running), you'll be presented to the installed components and you'll be able to select what you missed before.

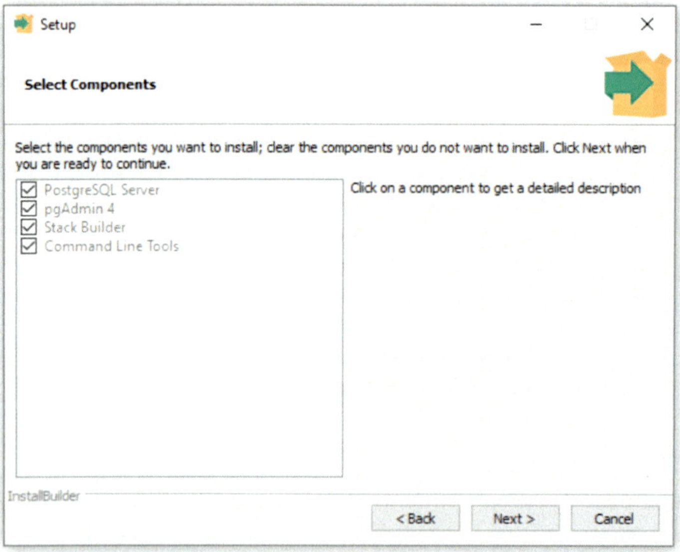

Select Components screen of a full upgrade installation shown on Windows

Click **Next** to proceed.

Upgrade installations will also present an extra **Existing installation** screen to inform you that the running instance might need to be restarted (depending of what will be installed / upgraded). This is important because it impacts potential running connections in the middle of their operations.

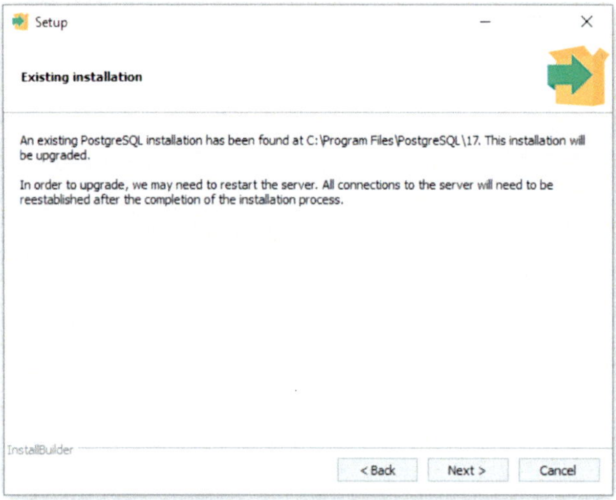

Select Components screen of a full upgrade installation shown on Windows

Click **Next** to proceed.

The next screens are exclusive to fresh installations and I'll be presenting screenshots only on macOS to avoid being too repetitive, as they are exactly the same as fresh installations on Windows.

Data Directory Configuration

The next screen asks for the data directory location, also known as the database cluster. This is where your physical database files will be stored.

Default locations: - **macOS**: `/Library/PostgreSQL/17/data` - **Windows**: `C:\Program Files\PostgreSQL\17/data`

For learning purposes, accept the default location within the main installation folder. In production environments, it's common to place the database cluster in a separate location for performance and management reasons.

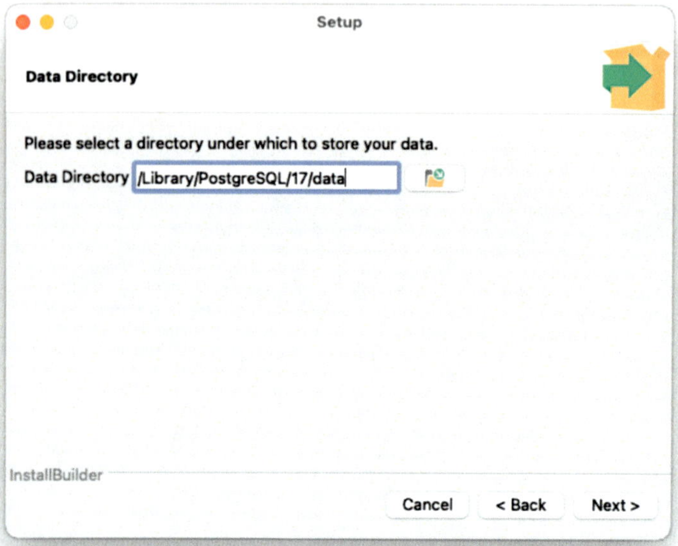

Data Directory selection screen on macOS

Database Superuser Password

You'll be prompted to create a password for the `postgres` superuser account. This is the master administrative account for your PostgreSQL installation.

Password Guidelines: - For learning environments on your local computer, choose an easy-to-remember password - For production environments, use strong passwords following your organization's security policies - Make it memorable—you'll need this password frequently for administrative tasks - Consider using a password manager

Remember: This password is required for: - Initial database connections - Administrative tasks - pgAdmin login - Command-line operations

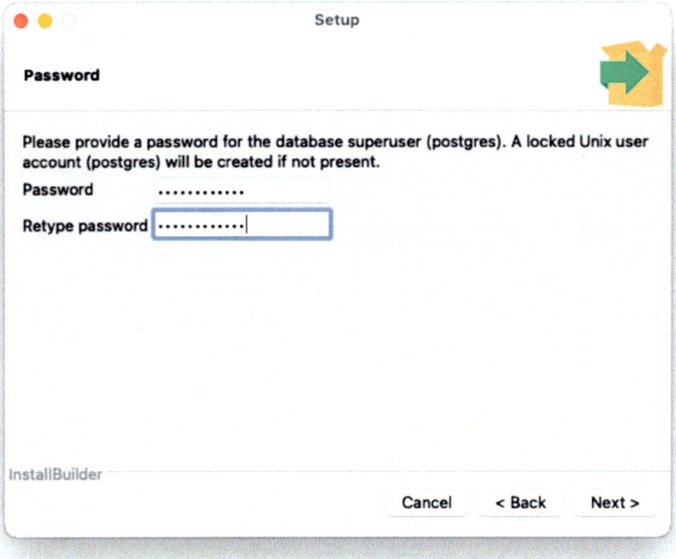

Password configuration dialog

Port Configuration

PostgreSQL needs a network port to listen for connections. The default port is 5432, which is well-known in the PostgreSQL community.

When to keep the default (5432): - Standard single-instance installation - No conflicting services - Following PostgreSQL conventions

When to change the port: - Another PostgreSQL instance is already running on 5432 - Your organization has specific port policies - You're running multiple PostgreSQL versions simultaneously

Remember your chosen port—you'll need to specify it in all connection strings if you change from the default.

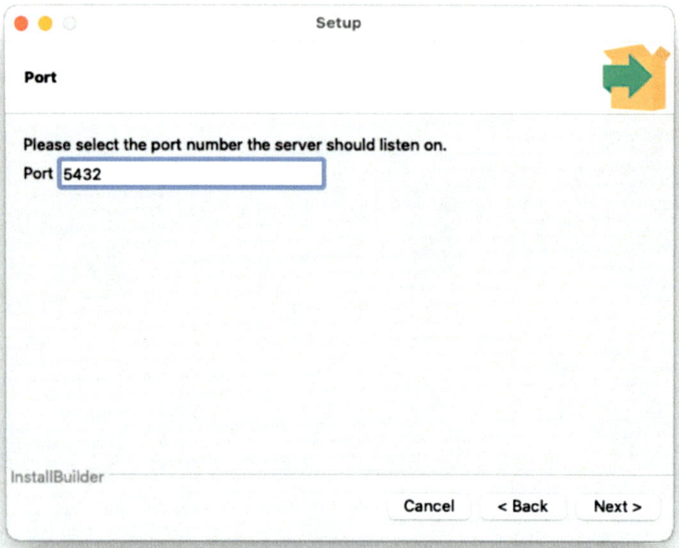

Port configuration screen

Locale and Regional Settings

The installer asks for locale settings that determine: - Text and character sorting and comparison rules - Date and time formatting - Number formatting and currency symbols - Other regional preferences

The locale format follows the pattern: language_country.encoding (for example, en_AU.UTF-8).

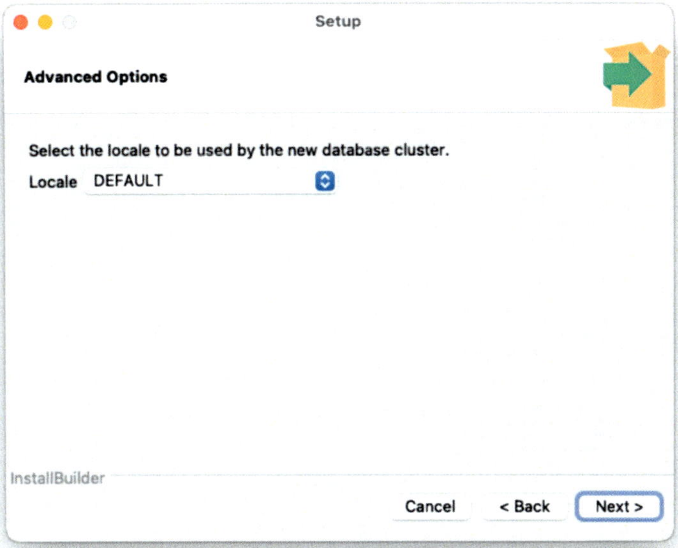

Locale configuration screen with DEFAULT

Recommendation: Choose your system's default locale unless you have specific requirements. You can create databases with different locales later if needed.

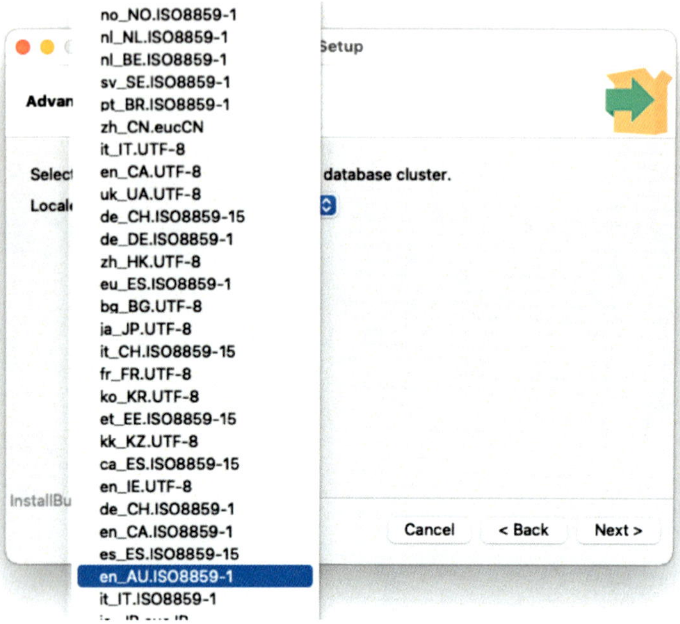

Locale configuration screen with another options

Completing the Installation

After configuring all settings, the installer displays a summary screen showing your choices.

Pre-Installation Summary

This screen appears on both platforms and shows identical information regardless of whether you're doing a fresh installation or an upgrade.

Review these settings carefully, paying particular attention to: - Installation directory - Selected components - Port number - Data directory location

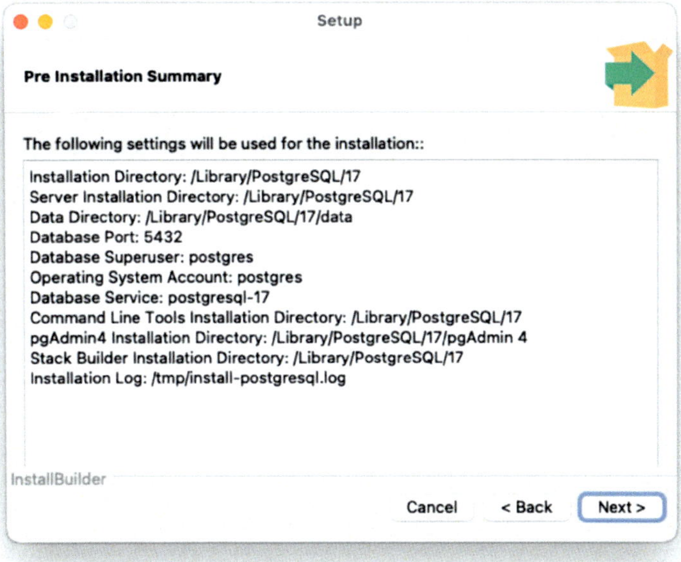

Installation summary dialog on macOS

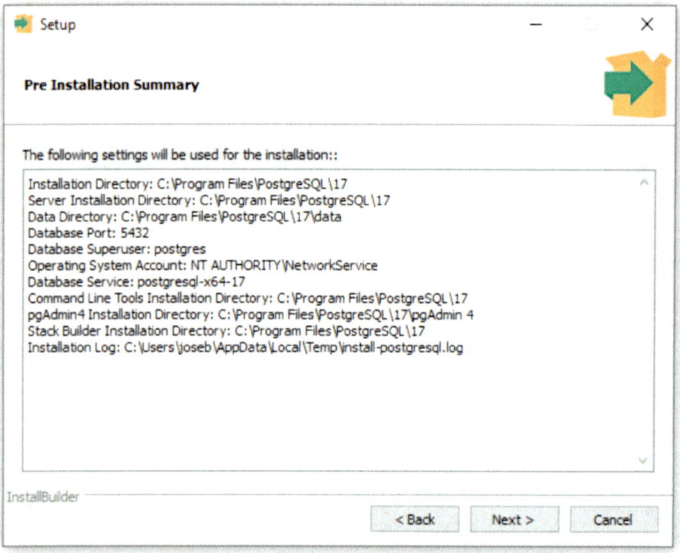

Installation summary dialog on Windows

Click **Next** to begin the installation. The process typically takes 2-5 minutes depending on your system.

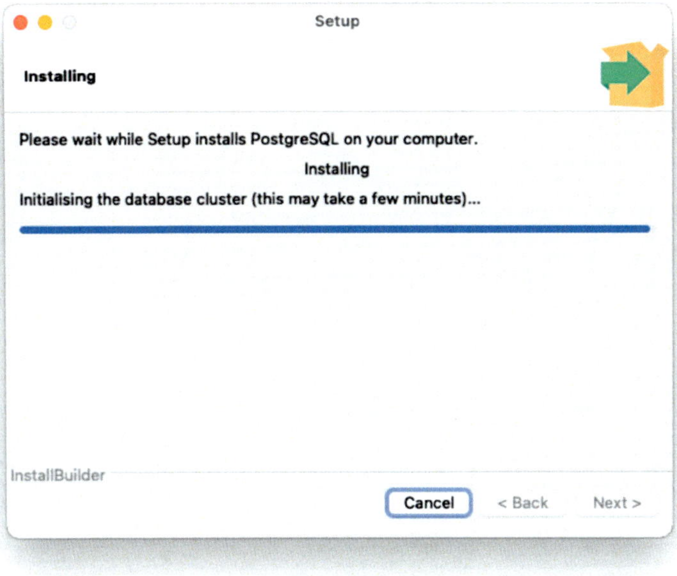

Installation Progress

Stack Builder Launch

After installation completes, you'll see a final screen offering to launch Stack Builder.

For first-time users: Uncheck the Stack Builder option. While it's useful for exploring the PostgreSQL ecosystem, it's not necessary for the core functionality.

For advanced users: Feel free to explore Stack Builder immediately to see available extensions and tools.

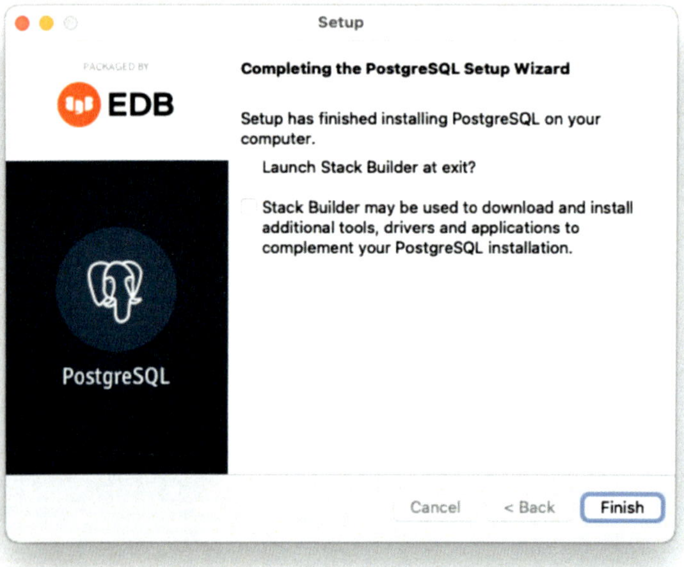

Completion screen with Stack Builder option

If you do choose to explore Stack Builder, you'll be presented to its Welcome screen with a dropdown where you select any configured server instance. From a typical local installation there will be just one option but Stack Builder is a PostgreSQL client that can even connect to remote servers.

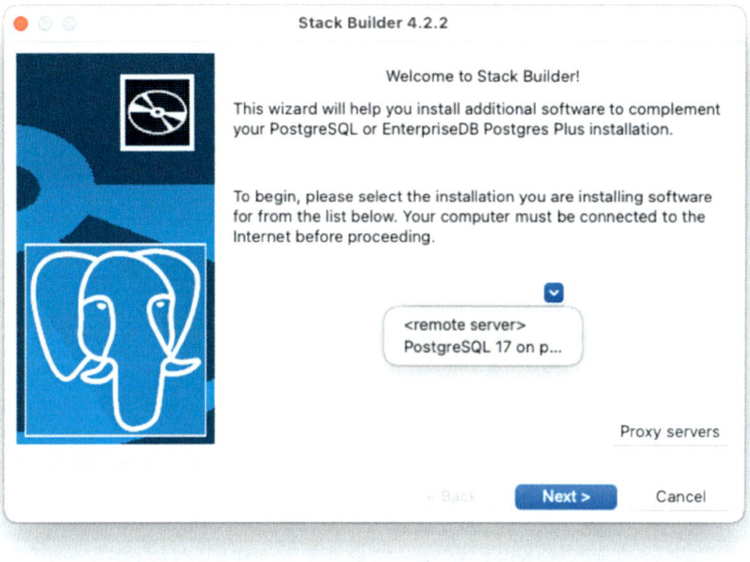

Completion screen with Stack Builder option

If you decide to go ahead and select your just installed PostgreSQL service and click on **Next**, you'll see it categorizes additional components into several groups:

- **Add-ons**: Including EDB language packs, pgAgent for job scheduling, and pgBouncer for connection pooling
- **Database drivers**: For various programming languages like .NET and JDBC drivers for Java
- **Spatial extensions**: Like PostGIS for geographic information systems (GIS) functionality

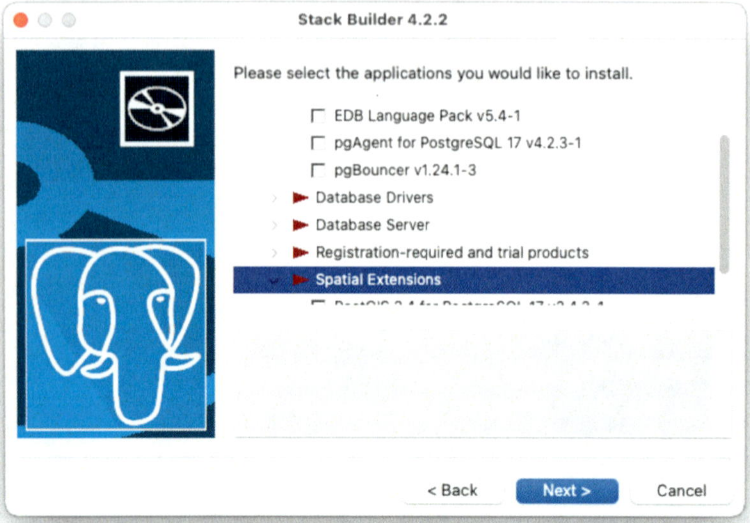

Completion screen with Stack Builder option

The Stack Builder downloads and presents current available packages, allowing you to select and install additional components as needed for your specific projects.

Hit cancel in case you don't want to install any extra component.

Initial Configuration and Verification

Congratulations! PostgreSQL is now installed on your system. Let's verify everything is working correctly and perform some initial configuration.

Default Installation Settings

Your PostgreSQL installation is now running with these default settings: - **Host**: localhost (127.0.0.1) - **Port**: 5432 (unless you changed it) - **Database**: postgres (default database) - **Username**: postgres (superuser account) - **Password**: what you set during installation

Service Management

PostgreSQL automatically starts as a system service on both platforms:

Windows: - The PostgreSQL service is installed and starts automatically with Windows - You can manage it through the Windows Services control panel (`services.msc`) - Look for "postgresql-x64-17" or similar service name

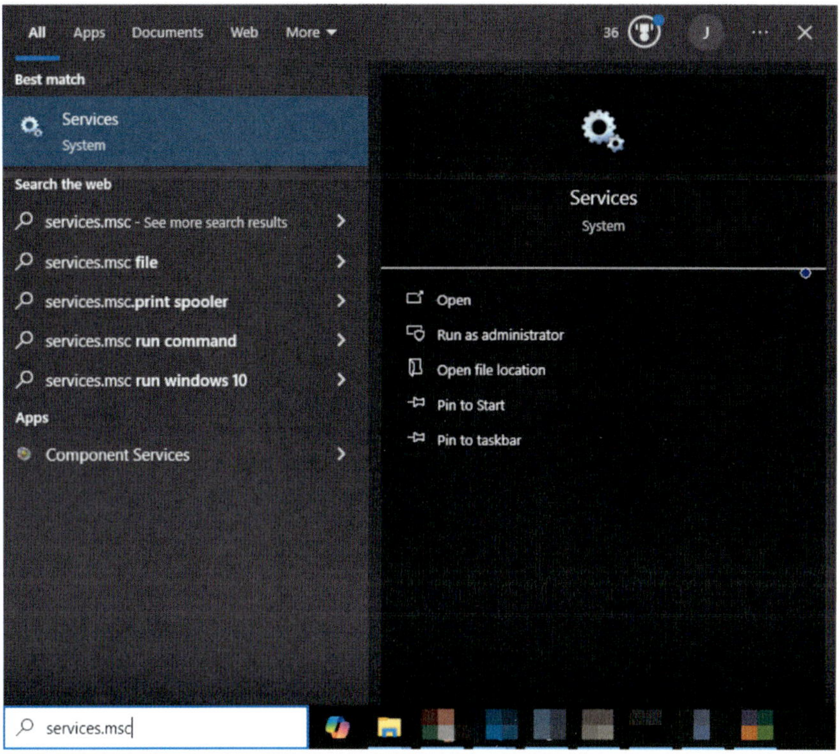

Running services.msc from Windows Start Menu

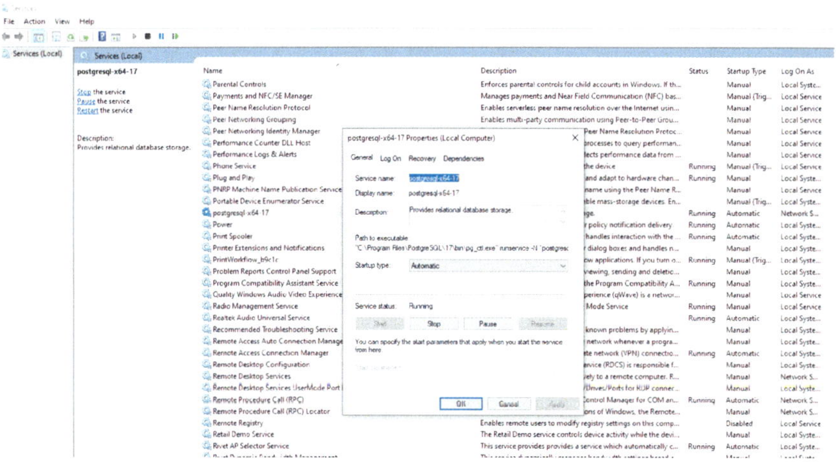

Finding PostgreSQL service running on Windows

macOS: - PostgreSQL starts automatically at system boot - Managed through system launch daemons (`launchd`) - No manual intervention typically required for startup

On next chapter, we'll cover this test as it's a serviced as command line operations.

Verification Using pgAdmin 4

Launch pgAdmin 4 from your applications: - **Windows**: Start Menu > pgAdmin 4 - **macOS**: Launchpad > pgAdmin 4

pgAdmin 4 will take a little while on an initial screen like this:

pgAdmin 4 starting screen

Then the initial screen will appear:

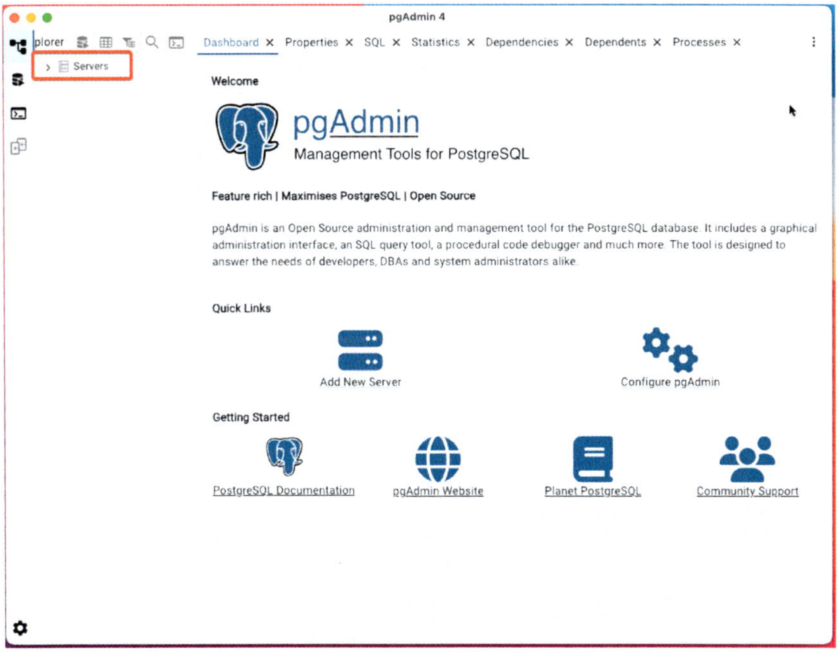

pgAdmin 4 initial screen

To verify your installation: 1. Expand the "Servers" tree in the left panel 2. You should see a "PostgreSQL 17" server entry 3. Right-click it and select "Connect Server" 4. Enter your postgres password when prompted 5. If successful, you'll see the server tree expand showing databases, users, and other objects

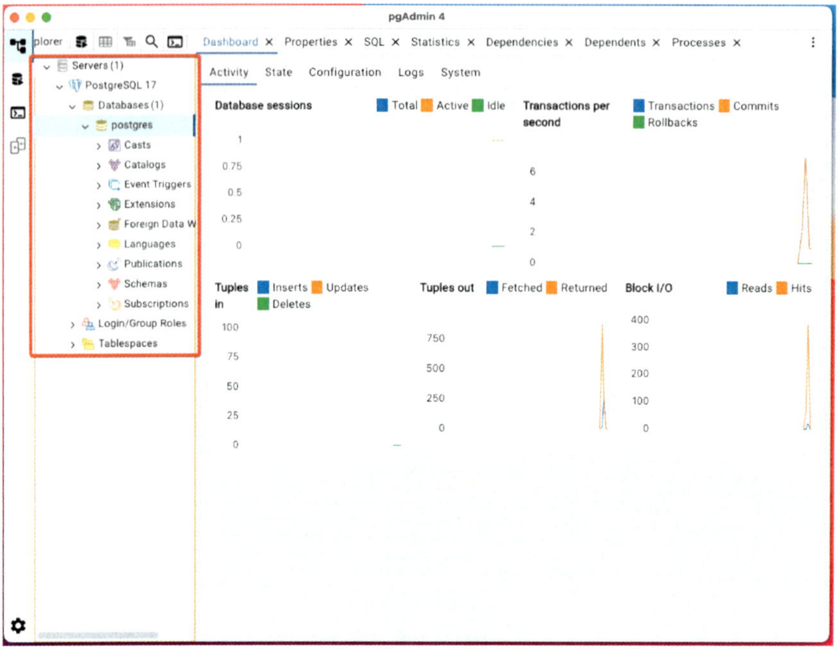

pgAdmin 4 expanded screen

Troubleshooting Common Issues

Let's address some common challenges you might encounter during or after installation.

Installation Issues

"Port already in use" Error If you see this error, another service is using port 5432: 1. Choose a different port during installation (e.g., 5433) 2. Or stop the conflicting service temporarily 3. Remember to use your chosen port in future connections

Permission Denied on macOS If macOS blocks the installation: 1. Go to System Preferences → Security & Privacy 2. Look for a message about the PostgreSQL installer 3. Click "Open Anyway" 4. Restart the installation process

Windows Defender or Antivirus Interference Some antivirus software may flag PostgreSQL: 1. Temporarily disable real-time scanning during installation 2. Add PostgreSQL installation directory to antivirus exclusions 3. Re-enable antivirus after successful installation

Post-Installation Issues

Cannot Connect to Database If pgAdmin or psql fails to connect: 1. Verify PostgreSQL service is running 2. Check if you're using the correct password 3. Confirm you're connecting to the right port 4. Try connecting to `localhost` instead of `127.0.0.1` or vice versa

Command Line Tools Not Found If `psql` and other commands aren't recognized:

Windows: Add PostgreSQL bin directory to your PATH environment variable: 1. Open System Properties → Advanced → Environment Variables 2. Edit the PATH variable 3. Add: `C:\Program Files\PostgreSQL\17\bin` 4. Restart Command Prompt

macOS: The installer shouldn't configure PATH automatically, so run the following command to get all

```
# It's always good to know upfront what is in the variable we'll change
echo $PATH
# If you don't see any reference to the PostgreSQL installation folder, run this
export PATH="/Library/PostgreSQL/17/bin:$PATH"
# Then confirm that PostgreSQL bin folder is the first path in $PATH
# Paths are separated by :
echo $PATH
```

Bonus: The Postgres.app for macOS

While the EnterpriseDB installer provides a comprehensive PostgreSQL installation experience, macOS users have access to an elegant alternative that embodies the Apple philosophy of simplicity: Postgres.app. This native macOS application offers a fundamentally different approach to PostgreSQL installation and management that some Mac users prefer.

Key Differences from EnterpriseDB Installer

Before diving into the installation process, it's important to understand how Postgres.app differs from the traditional installer approach we've covered:

Installation Philosophy: - **EnterpriseDB**: Traditional installer that requires admin privileges, goes through installation wizard steps, modifies system directories, and installs system services - **Postgres.app**: No installation required — it's a portable application that runs directly from wherever you place it (though the Applications folder is recommended)

System Integration: - **EnterpriseDB**: Installs PostgreSQL as a system service that starts automatically at boot, integrates deeply with the operating system - **Postgres.app**: Runs as a regular macOS app in your menu bar, starts and stops on demand when you open/close the application

Management Approach: - **EnterpriseDB**: Managed through system services, pgAdmin, and command-line tools - **Postgres.app**: Elegant menu bar interface with built-in server management, beautiful native macOS user interface

Components: - **EnterpriseDB**: Includes PostgreSQL server, pgAdmin 4, command-line tools, and Stack Builder - **Postgres.app**: Includes PostgreSQL server, command-line tools, and popular extensions like PostGIS, but no separate GUI tools

Accessing Postgres.app

Let's start from the same macOS download page we explored earlier. Navigate to postgresql.org/download/macosx, where you'll see the same page that offers both the EnterpriseDB installer and alternative installation methods.

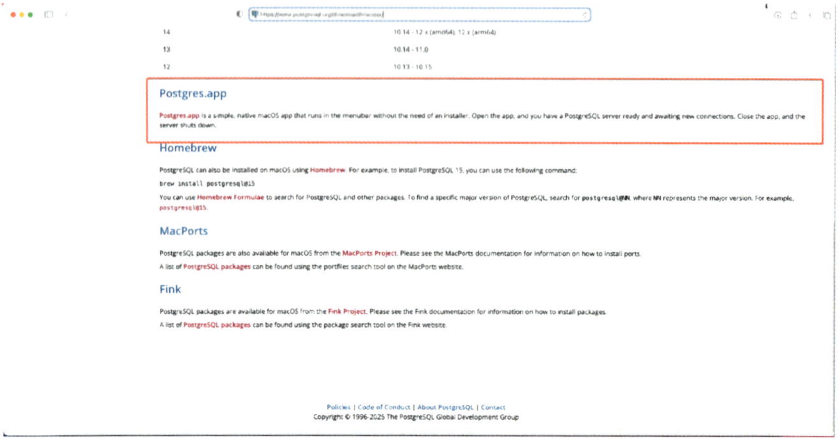

Postgres.app option on macOS download page

On this page, you'll notice a section mentioning "**Postgres.app** is a simple, native macOS app that runs in the menubar without the need of an installer." Click on this option or navigate directly to postgresapp.com.

Download and Setup Process

The Postgres.app website immediately showcases the simplicity of this approach with a clear three-step process:

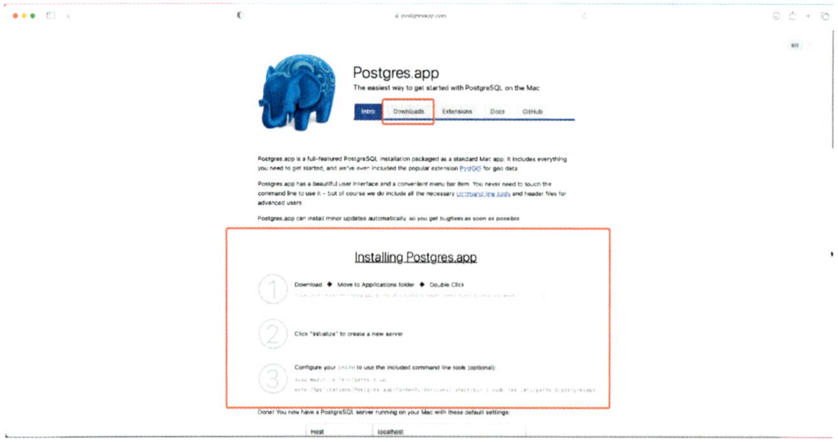

Postgres.app website showing its simple installation steps

Step 1: Download Click on **Downloads** from the main menu to go to the Downloads page, then click the Download button of the first panel to get the latest version of Postgres.app. The download is a standard

`.dmg` file, similar to any Mac application, typically around 119MB in size.

If requested, you can safely allow downloads from postgresapp.com.

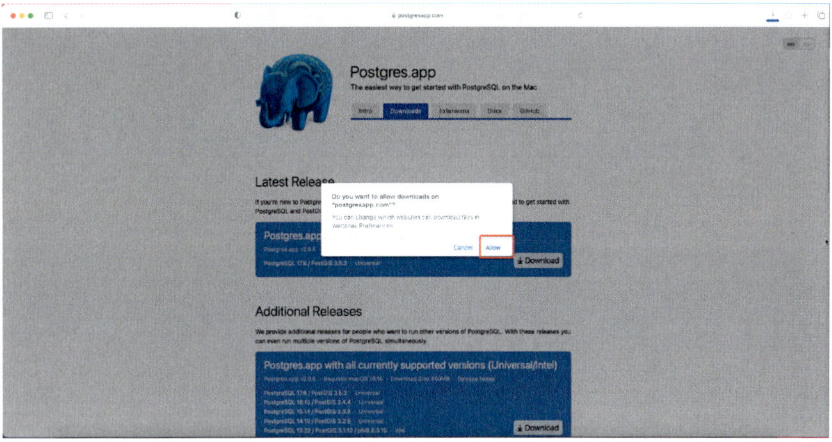

Postgres.app website showing its simple installation steps

Step 2: Move to Applications Folder Once downloaded, double-click the `.dmg` file to mount it. You'll see the familiar macOS installation interface with the Postgres.app icon and an Applications folder shortcut. Simply drag the Postgres.app icon to the Applications folder.

This is crucial: unlike many portable apps, Postgres.app should be placed in the Applications folder for optimal functionality. Some features may not work correctly if run from other locations.

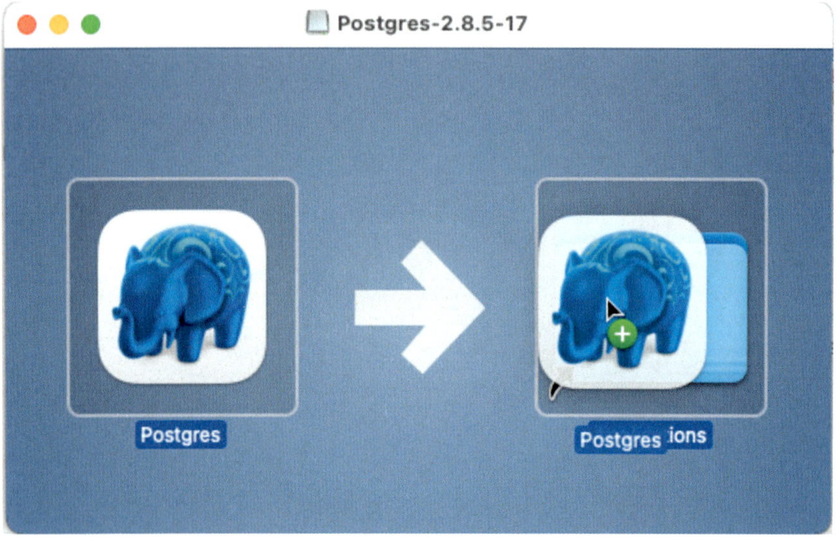

macOS drag-and-drop installation interface

Step 3: Double Click to Launch Navigate to your Applications folder and double-click Postgres.app to launch it. The first time you run it, macOS may display a security warning since it was downloaded from the internet. Click **Open** to proceed.

Initial Configuration

When Postgres.app launches for the first time, you'll see a clean, native macOS interface that's refreshingly different from typical database management tools.

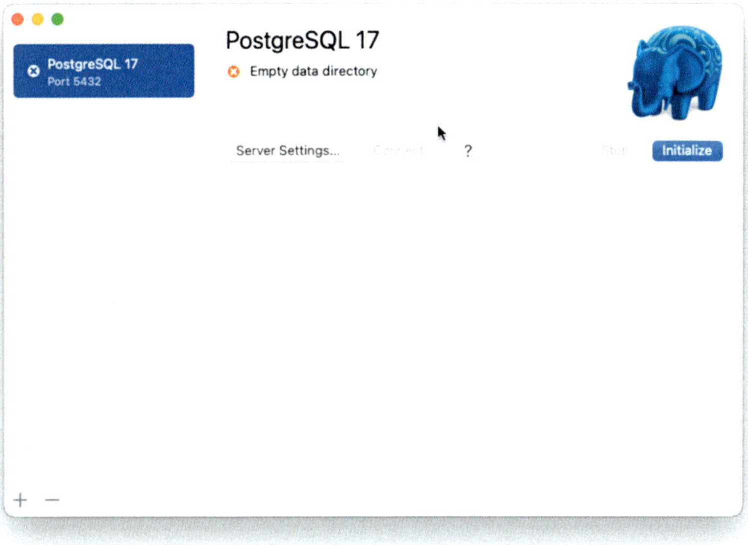

Postgres.app initial interface showing initialize button

Server Initialization: Click the **Initialize** button to create your first PostgreSQL server. This process: - Creates a new PostgreSQL data directory - Sets up the default databases - Configures the server with sensible defaults - Starts the PostgreSQL server automatically

Default Configuration: Once initialized, Postgres.app runs with these default settings: - **Host**: localhost - **Port**: 5432 - **User**: your macOS username (not "postgres") - **Database**: same as your username - **Password**: none required - **Connection URL**: postgresql://localhost/

This user-centric approach differs significantly from the EnterpriseDB installer, which creates a dedicated "postgres" superuser.

If you followed the EnterpriseDB installation before doing the Postgres.app, you'll face a **Port 5432 is already in use** error. And that's obviously because you now have two PostgreSQL server instances running and both are using the default TCP port.

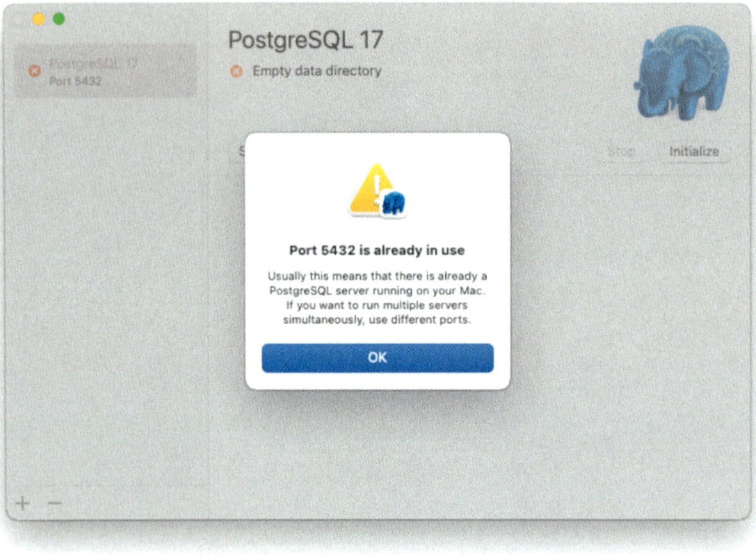

Postgres.app error when the port is already in use

By the way, a TCP port is exactly the way operating systems are capable of distributing clients connections to the right running service. If two servers could run on the same TCP port, there would be no way for the operating system to determine which one to deliver external connection attempts to.

To fix that, you can click on **Server Settings...**, then select another port, for example `5433` (of course it needs to be free too).

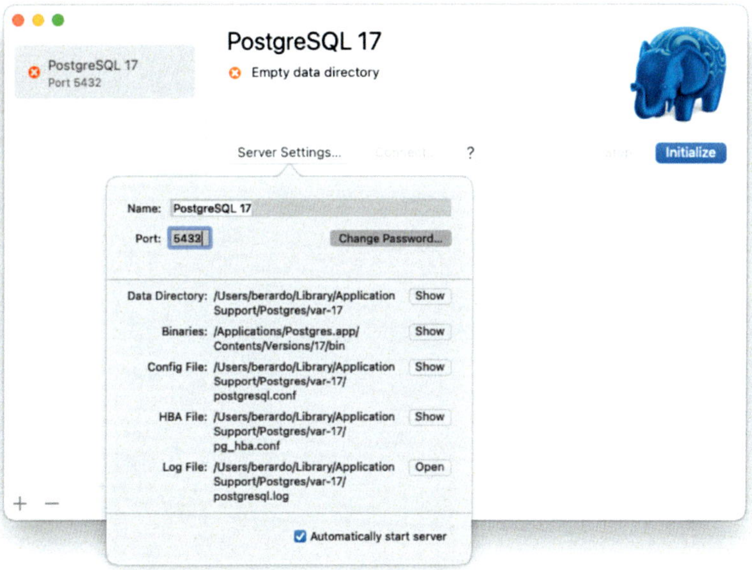

Postgres.app Server Settings popup

With the right port configured you should be able to **Initialize** your server and see three new running databases, as well as the green indicator that the server is running.

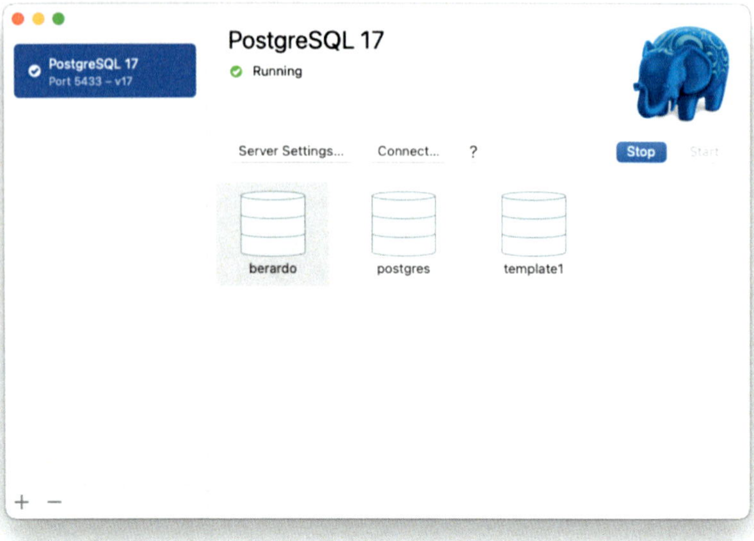

Postgres.app up and running screen

Menu Bar Integration

One of Postgres.app's most appealing features is its native macOS menu bar integration.

Menu Bar Features: - Click the elephant icon in your menu bar to access server controls - Start/stop the server with a single click - Quick access to server settings and information - Direct connection shortcuts to your databases

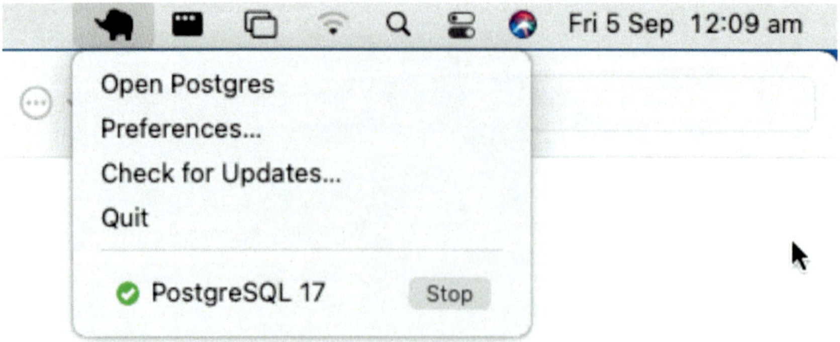

Postgres.app menu bar

Server Management: Unlike the EnterpriseDB installation, you have direct, visual control over your PostgreSQL server: - Green indicator when the server is running - Easy start/stop functionality - Quick access to server logs - Immediate feedback on server status

Command Line Integration

While Postgres.app provides an excellent GUI experience, it also includes full command-line tool support.

PATH Configuration (Optional): To use PostgreSQL command-line tools from Terminal, you can add Postgres.app's binaries to your PATH:

```
sudo mkdir -p /etc/paths.d &&
echo /Applications/Postgres.app/Contents/Versions/latest/bin | sudo tee /etc/paths.d/postgresapp
```

Direct Connection: Once configured, you can connect using standard PostgreSQL tools:

```
psql postgres
```

Database Access: For even easier access, double-click any database in the Postgres.app interface to open a Terminal window with psql already connected to that database.

Postgres.app up and running screen

Advantages of Postgres.app

Simplicity and Elegance: - No minimally complex installation wizard - No system service configuration No administrative privileges required for daily use - Clean, native macOS interface

Developer-Friendly: - Perfect for development and testing environments - Easy to start/stop servers as needed - Multiple PostgreSQL versions can coexist - Isolated from system-wide installations

Maintenance Benefits: - Automatic minor updates available - Easy to completely remove (just drag to trash) - Self-contained—doesn't scatter files across the system - No system service conflicts

When to Choose Postgres.app vs EnterpriseDB

Choose Postgres.app when: - You're developing on macOS and want a native experience - You prefer on-demand database server management - You want minimal system impact - You're working on personal or development projects - You appreciate elegant, simple interfaces

Choose EnterpriseDB installer when: - You need pgAdmin 4 included in the installation (although, you can also install it separately and connect to Postgres.app) - You want PostgreSQL to start automatically at boot - You're setting up a more permanent development environment - You need Stack Builder for additional extensions - You prefer traditional database administration approaches

Verification and Testing

Testing Your Installation: 1. Ensure Postgres.app is running (green indicator in menu bar) 2. Double-click the database with your username to open Terminal 3. Run a simple query to verify functionality:

```
SELECT version();
```

Expected Output: You should see PostgreSQL version information confirming that your Postgres.app installation is working correctly.

Terminal showing successful psql connection and version query

Integration with Development Workflow

Postgres.app excels in development environments where you need: - Quick server startup for development sessions - Easy switching between projects - Minimal resource usage when not actively developing - Clean development environment setup

Development Best Practices: - Start Postgres.app when beginning development work - Stop it when switching to other tasks to save system resources - Use the menu bar for quick status checks - Leverage the direct database connection features for rapid testing

With Postgres.app successfully set up, you have an elegant, native macOS PostgreSQL installation that embodies the best of Apple's design philosophy while providing full PostgreSQL functionality. This approach offers a refreshing alternative to traditional database installation methods and demonstrates how database tools can be both powerful and beautifully simple.

The choice between EnterpriseDB installer and Postgres.app ultimately comes down to your workflow preferences and requirements. Both provide excellent PostgreSQL experiences—the EnterpriseDB installer offers comprehensive tooling and traditional database management approaches, while Postgres.app delivers elegance, simplicity, and native macOS integration.

What's Next?

Excellent work! You now have a fully functional PostgreSQL installation with: - PostgreSQL 17 server running and ready for connections - pgAdmin 4 for graphical database management - Command-line tools for advanced operations - Stack Builder for future extensions and tools

Immediate Next Steps

1. **Bookmark Important Resources:**
    - Installation directory with command-line tools
    - PostgreSQL documentation: postgresql.org/docs
    - pgAdmin interface: pgadmin.org
2. **Explore pgAdmin Interface:** Spend a few minutes clicking through pgAdmin's interface. Notice the server tree, query editor, and dashboard.
3. **Create Your First Database:** In the last chapter, we'll cover some tools and configurations that common to any sort of installation process, then you'll see how to create your first database

Looking Ahead

The terminal-based installation methods covered in the next chapter explore command-line approaches that are essential for: - Server and production environments - Automation and scripting - Advanced configuration scenarios - Understanding PostgreSQL's underlying architecture

The visual installation approach covered in this chapter represents the most accessible entry point into PostgreSQL for development and learning environments. You've successfully installed one of the world's most advanced database systems, complete with professional-grade management tools. Whether you're building the next great application or learning database fundamentals, you now have the foundation to succeed.

Terminal Installation & Tools (Linux, macOS, Windows)

Welcome to our comprehensive exploration of PostgreSQL installation using terminal and command-line methods across all major platforms. After seeing the visual installation approach in our previous chapter, today we'll master the power of package managers - the preferred method for developers, system administrators, and anyone who appreciates automation.

By the end of this chapter, you'll understand not just *how* to install PostgreSQL via command line, but *why* this approach often provides superior control, easier updates, and better integration with your development workflow. Whether you're on Linux using `apt`, macOS with `brew`, or Windows with `chocolatey`, you'll see how similar principles apply across all platforms.

Package Manager Overview: The Universal Language

Despite running on different operating systems, modern package managers share remarkable similarities in both philosophy and usage patterns. Let's explore how `apt`, `brew`, and `chocolatey` approach the same fundamental challenges.

Package Managers on Linux

On previous chapter, we presented the official download page for Windows and macOS. So you certainly noticed the Linux icon on the downloads page.

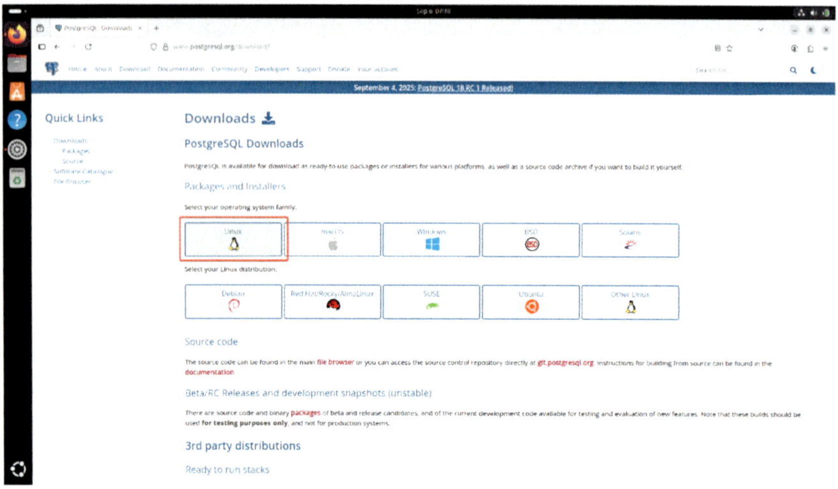

PostgreSQL official download page showing Linux downloads

However, contrary to Windows downloads that brings official information about the visual method, or macOS downloads that are officially covered on both visual and terminal methods, Linux installations are terminal only. Linux official download pages are instructions only. Installation files are either directly retrieved via package managers or from the source code page.

When you click on the Linux icon, you see a second line to choose from a few options:

- **Debian/Ubuntu**: Despite shown as separate download pages, their methods are equally based on Advanced Package Tool (apt). In this chapter, we'll go over this method running on Ubuntu
- **Red Hat/Fedora/Rocky Linux/AlmaLinux**: They are all grouped under the Red Hat company umbrella and uses the Yum package manager to install from RPM files
- **SUSE**: Also based on RPM files but from different repository

Linux: APT (Advanced Package Tool)

APT has been the foundation of Debian-based systems for over two decades. Originally developed for Debian Linux, it's now the standard package manager for Ubuntu and dozens of other distributions.

Core Philosophy: APT treats software as interconnected packages with dependencies that must be resolved systematically. It maintains a comprehensive database of available software and handles complex dependency chains automatically.

Key Characteristics: - System-level installation requiring `sudo` privileges - Automatic dependency resolution and conflict management - Integration with system services via `systemd` - Extensive repository ecosystem with thousands of packages

macOS: Homebrew

Homebrew revolutionized package management on macOS by bringing Unix-style package management to Apple's platform. Created in 2009, it's become the de facto standard for macOS developers.

Core Philosophy: "The missing package manager for macOS" - Homebrew fills the gap left by Apple's App Store model, providing command-line access to development tools and utilities.

Key Characteristics: - User-space installation (no `sudo` required for most operations) - Builds packages from source when needed - Integrates with macOS's `launchd` for service management - Massive ecosystem with over 6,000+ formulae

Windows: Chocolatey

Chocolatey brings Unix-style package management to Windows, addressing the platform's historical lack of a comprehensive package manager. Inspired by APT and Homebrew, it adapts these concepts to Windows conventions.

Core Philosophy: "The package manager for Windows" - Chocolatey treats Windows applications as packages that can be installed, updated, and managed programmatically.

Key Characteristics: - PowerShell-based with administrative privileges - Wraps existing Windows installers for package management - Integrates with Windows Services for background processes - Growing ecosystem with thousands of community packages

Universal Package Manager Patterns

Despite platform differences, all three follow similar patterns:

	Linux	macOS	Windows
Installation	`apt install postgresql`	`brew install postgresql`	`choco install postgresql`
Updates	`apt update && apt upgrade`	`brew update && brew upgrade`	`choco upgrade all`
Service Management	`systemctl`	`brew services`	Windows Services integration
Search	`apt search`	`brew search`	`choco search`

This consistency means mastering one package manager helps you understand them all.

Installation Commands: Cross-Platform Implementation

Now let's see how these principles translate into actual PostgreSQL installations. We'll execute each step across all three platforms simultaneously, highlighting both similarities and platform-specific considerations.

Step 1A: Package Manager Setup on Linux

Before installing PostgreSQL, we need to ensure our package managers are properly configured.

Linux (Ubuntu/Debian) - APT comes pre-installed on Debian-based systems, but we need to configure PostgreSQL's official repository for the latest versions:

```
# Update package database
sudo apt update

# Install repository management package
sudo apt install -y postgresql-common

# Add PostgreSQL Apt repository using the official helper script
sudo /usr/share/postgresql-common/pgdg/apt.postgresql.org.sh
```

PostgreSQL repository added to Apt package manager

Alternative Manual Configuration (if needed):

If the automated script doesn't work or you want to understand what the automated process does behind the scenes, follow these steps:

```
# Install required dependencies for HTTPS and certificates
sudo apt install -y curl ca-certificates

# Create directory for repository files
sudo install -d /usr/share/postgresql-common/pgdg

# Download PostgreSQL repository signing key
sudo curl -o /usr/share/postgresql-common/pgdg/apt.postgresql.org.asc --fail https://www.postgresql.org/media/keys/ACCC4CF8.asc
```

```
# Create repository configuration file
. /etc/os-release
sudo sh -c 'echo "deb [signed-by=/usr/share/postgresql-common/pgdg/apt.postgresql.org.asc] https://apt.postgresql.org/pub/repos/apt $VERSION_CODENAME-pgdg main" > /etc/apt/sources.list.d/pgdg.list'

# Update package database with new repository
sudo apt update
```

Why These Steps Matter:

- The **helper script** (`apt.postgresql.org.sh`) is the official PostgreSQL method for adding their repository, ensuring you get the latest supported versions rather than potentially outdated packages from Ubuntu's default repositories
- The **signing key** ensures package authenticity and security - APT verifies that downloaded packages haven't been tampered with
- **Repository configuration** tells APT exactly where to find PostgreSQL packages and which key to use for verification
- Using `$VERSION_CODENAME` ensures the correct packages for your specific Ubuntu release (like **jammy** for 22.04 or **noble** for 24.04)

This approach follows the current official PostgreSQL documentation while providing the educational context that makes each step clear and meaningful for readers learning the process.

Step 1B: Package Manager Setup on macOS (Homebrew)

Homebrew requires initial installation if not already present.

1. Install Homebrew (if not already installed):

```
/bin/bash -c "$(curl -fsSL https://raw.githubusercontent.com/Homebrew/install/HEAD/install.sh)"
```

This command downloads and runs the official Homebrew installation script.

2. Configure your shell environment:

Add Homebrew to your shell profile to ensure the `brew` command is accessible:

```
# For zsh shell users (default on macOS)
echo 'eval "$(/opt/homebrew/bin/brew shellenv)"' >> ~/.zprofile

# Apply changes immediately
eval "$(/opt/homebrew/bin/brew shellenv)"
```

Note: Depending on your shell (bash, zsh, etc.) you might need to add this to a different file such as `.bash_profile` or `.bashrc`.

3. Update Homebrew package metadata:

```
brew update
```

This fetches the latest list of available packages and updates Homebrew metadata.

Why These Steps Matter:

- The **Homebrew installation script** automatically detects your macOS version and architecture (Intel vs Apple Silicon) to configure the appropriate installation paths
- **Shell environment configuration** ensures the `brew` command is available in all new terminal sessions by adding it to your PATH
- **Package metadata updates** refresh Homebrew's local database to ensure you get the latest versions and security updates

Important Notes for Homebrew:

- Homebrew installs packages to user-owned directories, so you do *not* need to run commands with `sudo`
- This reduces permission issues and makes Homebrew user-friendly
- It integrates with macOS's native `launchd` system for managing background services like PostgreSQL

Step 1C: Package Manager Setup on Windows (Chocolatey)

Chocolatey is a package manager for Windows that operates through PowerShell and Windows' package management frameworks.

> **Important:** Most Chocolatey commands require you to run PowerShell with Administrator privileges (`Run as Administrator`). This is **necessary** because installing or managing software system-wide involves modifying protected areas which require elevated permissions.

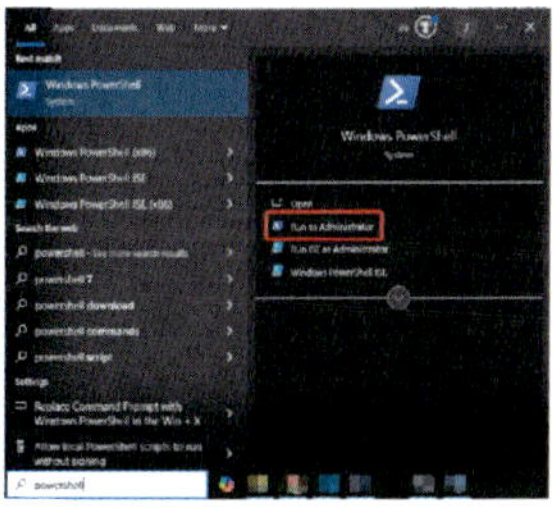

Run Powershell as Administrator

1. Check current PowerShell execution policy:

```
Get-ExecutionPolicy
```

This command shows the current script execution policy. If it returns Restricted, you will need to temporarily bypass it to allow script execution.

2. Temporarily bypass script restrictions for this session:

```
Set-ExecutionPolicy Bypass -Scope Process -Force
```

This change is temporary and applies only to the current PowerShell session without modifying system-wide security settings.

3. Install Chocolatey:

Execute the official installation command:

```
Set-ExecutionPolicy Bypass -Scope Process -Force; [System.Net.ServicePointManager]::SecurityProtocol = [System.Net.ServicePointManager]::SecurityProtocol -bor 3072; iex ((New-Object System.Net.WebClient).DownloadString('https://community.chocolatey.org/install.ps1'))
```

4. Verify installation:

Close and reopen your PowerShell *as Administrator*, then run:

```
choco --version
```

You should see the Chocolatey version installed.

Why These Steps Matter:

- **Execution policy checks** ensure your system allows running the installation scripts while maintaining security
- **TLS security protocol configuration** ensures secure downloads from Chocolatey's servers
- **Administrative privileges verification** confirms you have the necessary permissions for system-wide package management

Administrative Rights - Necessary vs Recommended:

- Running PowerShell as Administrator is **necessary** (not just recommended) for:
  - Installing Chocolatey itself
  - Installing packages that modify system directories
  - Managing Windows services (like PostgreSQL)
  - Updating system PATH environment variables
- Chocolatey installs software system-wide and manages Windows services, which requires elevated permissions by Windows design
- If you run Chocolatey commands without administrative rights, installations will fail with permission errors
- You may need to add the Chocolatey path (`C:\ProgramData\chocolatey\bin`) to your system PATH environment variables if `choco` is not recognized after installation

This approach ensures you understand both the technical steps and the security requirements for successful PostgreSQL installation on Windows.

Step 2: PostgreSQL Installation

With package managers configured, installing PostgreSQL becomes remarkably similar across platforms.

Linux (APT):

```
# Install PostgreSQL server and contributed packages
sudo apt install postgresql postgresql-contrib

# Installation automatically creates 'postgres' system user
# Service starts automatically via systemd
```

macOS (Homebrew):

```
# Install PostgreSQL (latest stable version)
brew install postgresql

# Or install specific version
brew install postgresql@17

# Homebrew doesn't auto-start services by default
```

Windows (Chocolatey):

```
# Install PostgreSQL (includes server, client tools, pgAdmin)
choco install postgresql

# Installation creates Windows service
# Generates random password for 'postgres' user (note this!)
```

Cross-Platform Observations: Notice how each package manager handles PostgreSQL slightly differently: - **APT**: Immediately starts service and integrates with system users - **Homebrew**: Installs but requires manual service start - **Chocolatey**: Creates Windows service but generates a random password

These differences reflect each platform's conventions, but the core installation process remains conceptually identical.

Step 3: Verification of Installation

Let's verify our installations work correctly across all platforms.

Linux:

```
# Check PostgreSQL version
psql --version

# Verify service status
sudo systemctl status postgresql

# Test connection as postgres user
sudo -u postgres psql -c "SELECT version();"
```

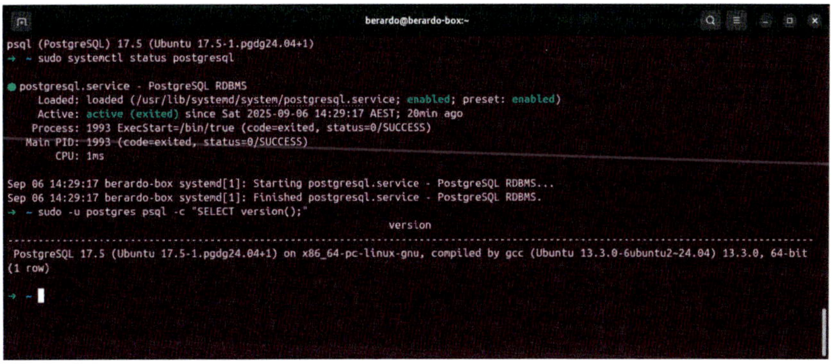

PostgreSQL up and running on Ubuntu

macOS:

```
# Check PostgreSQL version
psql --version
```

```
# Check if service is running (should show 'none' if not started)
brew services list | grep postgresql

# Test connection (may need to start service first)
psql postgres -c "SELECT version();"
```

Windows:

```
# Check PostgreSQL version
psql --version

# Check Windows service status
Get-Service -Name postgresql*

# Test connection with postgres user
psql -U postgres -c "SELECT version();"
```

If any of these commands fail, don't worry - we'll address common issues in our troubleshooting section.

Service Management: Platform-Specific Approaches

Managing PostgreSQL as a background service differs significantly across platforms, reflecting each operating system's service management philosophy.

Linux: systemd Integration

Modern Linux distributions use `systemd` for service management, providing robust service control with automatic startup, restart on failure, and comprehensive logging.

```
# Start PostgreSQL service
sudo systemctl start postgresql

# Enable automatic startup at boot
sudo systemctl enable postgresql

# Check service status (detailed information)
sudo systemctl status postgresql

# View service logs
sudo journalctl -u postgresql -f

# Restart service (useful after configuration changes)
sudo systemctl restart postgresql

# Stop service
sudo systemctl stop postgresql
```

systemd Advantages: Dependency management, automatic restart on failure, resource limitation, and comprehensive logging make systemd ideal for production environments.

macOS: Homebrew Services with launchd

Homebrew abstracts macOS's `launchd` service management system, providing a user-friendly interface while leveraging Apple's robust service infrastructure.

```
# Start PostgreSQL service and enable auto-start
brew services start postgresql

# Start service without auto-start (current session only)
brew services run postgresql

# Check service status
brew services list

# Stop service and disable auto-start
brew services stop postgresql

# Restart service
brew services restart postgresql

# View service information
brew services info postgresql
```

Homebrew Services Integration: The `brew services` command seamlessly integrates with macOS's `launchd`, creating appropriate `.plist` files and managing service lifecycle without requiring manual `launchctl` interaction.

Windows: Service Manager Integration

Chocolatey installations integrate with Windows Services, providing both GUI and command-line management options.

```
# Check PostgreSQL service status
Get-Service -Name postgresql*

# Start PostgreSQL service
Start-Service -Name postgresql-x64-17

# Stop PostgreSQL service
Stop-Service -Name postgresql-x64-17

# Set service to start automatically
Set-Service -Name postgresql-x64-17 -StartupType Automatic
```

```
# View service configuration
Get-WmiObject -Class Win32_Service -Filter "Name='postgresql-x64-17'"
```

Windows Service Integration: PostgreSQL integrates with Windows Services infrastructure, allowing management through Services.msc GUI or PowerShell commands. The service runs under a dedicated PostgreSQL user account created during installation.

Cross-Platform Service Management Patterns

Despite different implementations, service management follows consistent patterns:

Start/Stop: All platforms provide immediate service control **Auto-Start**: All systems support automatic startup at boot/login **Status Checking**: Each platform offers status verification commands **Logging**: Comprehensive logging helps troubleshoot service issues

Understanding these patterns helps you manage PostgreSQL services regardless of platform.

Configuration: Post-Installation Essentials

After successful installation, several configuration steps ensure optimal PostgreSQL operation across all platforms. While specific file locations vary, the configuration concepts remain consistent.

PostgreSQL Configuration Files

PostgreSQL uses three primary configuration files that work identically across platforms:

postgresql.conf: Main server configuration (memory, connections, performance) **pg_hba.conf**: Client authentication rules (who can connect and how) **pg_ident.conf**: User name mapping (optional, for advanced authentication)

OS	Configuration File Locations
Linux	`/etc/postgresql/[version]/main/` or `/var/lib/pgsql/data/`
macOS (Homebrew)	`/opt/homebrew/var/postgresql@17/` or `/usr/local/var/postgresql/`
Windows	`C:\Program Files\PostgreSQL\17\data\`

Essential Configuration Steps

Let's perform crucial configuration tasks across all platforms.

Step 1: Set postgres User Password

The `postgres` superuser needs a secure password for administrative tasks.

Linux:

```
# Switch to postgres user and set password
sudo -u postgres psql
```

```
\password postgres
# Enter new password when prompted
\q
```

macOS:

```
# Connect as current user to postgres database
psql postgres
\password postgres
# Enter new password when prompted
\q
```

Windows:

```
# Connect using the random password from installation
psql -U postgres
\password postgres
# Enter new password when prompted
\q
```

Step 2: Configure Connection Settings

Basic connection configuration ensures PostgreSQL accepts connections properly.

Edit postgresql.conf (adjust path for your platform):

```
# Linux
sudo vim /etc/postgresql/17/main/postgresql.conf

# macOS
nano /opt/homebrew/var/postgresql@17/postgresql.conf

# Windows (as Administrator)
notepad "C:\Program Files\PostgreSQL\17\data\postgresql.conf"
```

Key Settings to Verify:

```
# Connection Settings
listen_addresses = 'localhost'          # For local connections only
port = 5432                             # Standard PostgreSQL port
max_connections = 100                   # Adjust based on needs

# Memory Settings (adjust based on system resources)
shared_buffers = 128MB                  # Start with 25% of RAM
work_mem = 4MB                          # Per-query memory
maintenance_work_mem = 64MB             # Maintenance operations
```

Step 3: Configure Authentication

Edit `pg_hba.conf` to set authentication rules:

```
# Linux
sudo nano /etc/postgresql/17/main/pg_hba.conf

# macOS
nano /opt/homebrew/var/postgresql@17/pg_hba.conf

# Windows (as Administrator)
notepad "C:\Program Files\PostgreSQL\17\data\pg_hba.conf"
```

Simple Authentication Settings (unsafe for production environments):

```
# TYPE  DATABASE        USER            ADDRESS                 METHOD

# "local" is for Unix domain socket connections only
local   all             all                                     peer
# IPv4 local connections:
host    all             all             127.0.0.1/32            trust
# IPv6 local connections:
host    all             all             ::1/128                 scram-sha-256
```

Step 4: Apply Configuration Changes

After modifying configuration files, restart PostgreSQL to apply changes:

Linux:

```
# To fully restart the service
sudo systemctl restart postgresql
# To see current running status
sudo systemctl status postgresql
```

If you only change pg_hba.conf, a `reload` command is sufficient:

```
sudo systemctl reload postgresql
```

This command sends a `SIGHUP` signal to the PostgreSQL server process, instructing it to re-read its configuration files, including pg_hba.conf. This allows the changes you made to take effect without requiring a full restart of the PostgreSQL server. This means existing connections will not be terminated, and the new authentication rules will take effect for new connections.

If your changes involve postgresql.conf file they might require a full restart.

macOS:

```
# To fully restart the service
brew services restart postgresql
# To see current running status
brew services list | grep postgresql
```

Similar to Linux, `brew reload` can also be used to force the server to re-apply (potentially changed) configurations, particularly those made on pg_hba.conf:

```
brew services reload postgresql
```

Windows: Unfortunately, Windows does not have a native "reload" command equivalent to Unix/Linux systems. On the other hand, you have two options to restart the postgres service. One we covered on visual installation, which is using `services.msc`. As Chocolatey also installs PostgreSQL as a system service, you can always manage it through this visual interface.

Alternatively, you can run it via Powershell:

```
# To fully restart the service
Restart-Service postgresql-x64-17
# To see current running status
Get-Service postgresql-x64-17
```

Creating Your First Database

With PostgreSQL properly configured, let's create a database for testing:

All Platforms:

```
# Connect to PostgreSQL
psql -U postgres

# Create a test database
CREATE DATABASE testdb;

# Create a regular user (recommended for applications)
CREATE USER myuser WITH PASSWORD 'mypassword';

# Grant privileges
GRANT ALL PRIVILEGES ON DATABASE testdb TO myuser;

# List databases to verify
\l

# Exit
\q
```

Test the new user connection:

```
psql -U myuser -d testdb -h localhost
```

If this connection succeeds, your PostgreSQL installation is fully functional!

Verification: Testing Your Installation

A systematic verification process ensures your PostgreSQL installation works correctly across all platforms. Let's test core functionality step by step.

Basic Connectivity Tests

Test 1: Version Verification

```
# Should work on all platforms
psql --version
```

Expected Output: `psql (PostgreSQL) 17.x`

Test 2: Server Connection

```
# Linux
sudo -u postgres psql -c "SELECT version();"

# macOS
psql postgres -c "SELECT version();"

# Windows
psql -U postgres -c "SELECT version();"
```

Expected Output: Full PostgreSQL version information with system details

Test 3: Service Status Check

```
# Linux
sudo systemctl is-active postgresql

# macOS
brew services list | grep postgresql | awk '{print $2}'

# Windows (PowerShell)
(Get-Service postgresql-x64-17).Status
```

Expected Output: "active" (Linux), "started" (macOS), or "Running" (Windows)

Functional Database Tests

Test 4: Database Operations

```
-- Connect to PostgreSQL (using appropriate method for your platform)
psql -U postgres

-- Create test database
CREATE DATABASE verification_test;

-- Connect to test database
\c verification_test

-- Create test table
CREATE TABLE test_table (
```

```
    id SERIAL PRIMARY KEY,
    name VARCHAR(50),
    created_at TIMESTAMP DEFAULT CURRENT_TIMESTAMP
);

-- Insert test data
INSERT INTO test_table (name) VALUES ('Test Entry 1'), ('Test Entry 2');

-- Query test data
SELECT * FROM test_table;

-- Clean up
DROP TABLE test_table;
\c postgres
DROP DATABASE verification_test;
\q
```

If all these operations complete successfully, your PostgreSQL installation is working correctly!

Performance and Configuration Tests

Test 5: Configuration Verification

```
# Connect and check key settings
psql -U postgres -c "SHOW shared_buffers;"
psql -U postgres -c "SHOW max_connections;"
psql -U postgres -c "SHOW port;"
```

Test 6: User Authentication

```
# Test password authentication (should prompt for password)
psql -U postgres -h localhost
```

Troubleshooting Common Issues

Even with careful installation, you might encounter platform-specific challenges. Let's address the most common issues systematically.

Package Manager Issues

Linux: APT Repository Problems

```
# Issue: GPG key verification failed
# Solution: Re-import PostgreSQL signing key
sudo apt-key adv --keyserver keyserver.ubuntu.com --recv-keys ACCC4CF8

# Issue: Repository not found
# Solution: Verify repository configuration
```

```
cat /etc/apt/sources.list.d/pgdg.list

# Issue: Package conflicts
# Solution: Clean package cache
sudo apt clean && sudo apt update
```

macOS: Homebrew Installation Problems

```
# Issue: Homebrew not in PATH
# Solution: Add Homebrew to shell profile
echo 'eval "$(/opt/homebrew/bin/brew shellenv)"' >> ~/.zprofile

# Issue: Permission denied
# Solution: Fix Homebrew ownership
sudo chown -R $(whoami) /opt/homebrew/*

# Issue: Conflicting PostgreSQL installations
# Solution: Uninstall other versions
brew uninstall --ignore-dependencies postgresql
```

Windows: Chocolatey PowerShell Issues

```
# Issue: Execution policy restricted
# Solution: Set appropriate execution policy
Set-ExecutionPolicy RemoteSigned -Scope CurrentUser

# Issue: Admin privileges required
# Solution: Run PowerShell as Administrator
# Right-click PowerShell → "Run as Administrator"

# Issue: Chocolatey not recognized
# Solution: Restart PowerShell or add to PATH
$env:Path += ";C:\ProgramData\chocolatey\bin"
```

Service Management Issues

Linux: systemd Service Problems

```
# Issue: Service won't start
# Check service logs
sudo journalctl -u postgresql -n 50

# Common causes and solutions:
# 1. Port already in use
sudo lsof -i :5432
sudo systemctl stop conflicting-service
```

```
# 2. Data directory permissions
sudo chown -R postgres:postgres /var/lib/postgresql/
sudo chmod 700 /var/lib/postgresql/*/main

# 3. Configuration file errors
sudo -u postgres /usr/lib/postgresql/17/bin/postgres -D /var/lib/postgresql/17/main --check
```

macOS: Homebrew Services Issues

```
# Issue: Service won't start
# Check PostgreSQL logs
tail -f /opt/homebrew/var/log/postgresql@17.log

# Common solutions:
# 1. Data directory initialization
/opt/homebrew/bin/initdb --locale=C -E UTF-8 /opt/homebrew/var/postgresql@17

# 2. Port conflicts
lsof -i :5432

# 3. Permission issues
chmod 700 /opt/homebrew/var/postgresql@17
```

Windows: Service Configuration Issues

```
# Issue: Service won't start
# Check Windows Event Logs
Get-EventLog -LogName Application -Source PostgreSQL -Newest 10

# Common solutions:
# 1. Service account permissions
# Run Services.msc → PostgreSQL → Properties → Log On tab

# 2. Data directory access
# Ensure PostgreSQL service account has full control over data directory

# 3. Firewall blocking connections
New-NetFirewallRule -DisplayName "PostgreSQL" -Direction Inbound -Port 5432 -Protocol TCP -Action Allow
```

Connection and Authentication Issues

Cross-Platform Connection Problems:

Issue: "psql: command not found"

```
# Linux: Add PostgreSQL binaries to PATH
export PATH=$PATH:/usr/lib/postgresql/17/bin
```

```
# macOS: Use full path or add to PATH
export PATH="/opt/homebrew/bin:$PATH"

# Windows: Add PostgreSQL bin to system PATH
# System Properties → Environment Variables → PATH → Add "C:\Program Files\PostgreSQL\17\bin"
```

Issue: "connection refused"

```
# Check if PostgreSQL is running
# Linux
sudo systemctl status postgresql

# macOS
brew services list | grep postgresql

# Windows
Get-Service postgresql*

# Check if port is blocked
telnet localhost 5432
```

Issue: Authentication failed

```
# Verify pg_hba.conf settings
# Make sure authentication method matches your setup

# Reset postgres password
sudo -u postgres psql
\password postgres

# Test with specific authentication method
psql -U postgres -h localhost
```

Platform-Specific Best Practices

As we conclude our terminal installation journey, let's highlight best practices specific to each platform that will serve you well in production environments.

Linux Best Practices

System Integration: - Always use the official PostgreSQL APT repository for latest versions and security updates - Configure `systemd` service files for custom startup requirements - Use `logrotate` to manage PostgreSQL log files automatically - Set appropriate `ulimit` values for the postgres user

Security Considerations:

```
# Create dedicated backup user
sudo adduser pgbackup
```

```
sudo usermod -aG postgres pgbackup

# Configure firewall (if needed)
sudo ufw allow from 10.0.0.0/8 to any port 5432

# Set up SSL certificates
sudo mkdir /etc/postgresql/17/main/ssl
sudo chown postgres:postgres /etc/postgresql/17/main/ssl
```

macOS Best Practices

Development Environment Optimization: - Use Homebrew services for automatic startup management - Configure shell environment for PostgreSQL tools - Integrate with macOS Keychain for password management

Multiple Version Management:

```
# Install specific versions for testing
brew install postgresql@15 postgresql@16 postgresql@17

# Switch between versions
brew unlink postgresql@16
brew link postgresql@17

# Use different ports for parallel installations
```

Windows Best Practices

Enterprise Environment Integration: - Configure Windows authentication integration for enterprise environments - Use Windows Task Scheduler for automated maintenance tasks - Integrate with Windows Performance Monitor for system monitoring

PowerShell Automation:

```
# Create PowerShell profile for PostgreSQL tools
New-Item -Type File -Path $PROFILE -Force
Add-Content $PROFILE '$env:PATH += ";C:\Program Files\PostgreSQL\17\bin"'

# Configure Windows Firewall rules
New-NetFirewallRule -DisplayName "PostgreSQL-In" -Direction Inbound -Port 5432 -Protocol TCP -Action Allow
```

Next Steps: Building on Your Foundation

Congratulations! You've successfully mastered PostgreSQL installation across all major platforms using terminal and command-line methods. This foundation prepares you for the advanced topics ahead in our course.

Immediate Next Steps

1. **Practice Service Management**: Try starting, stopping, and restarting PostgreSQL on your platform to build confidence with service commands.
2. **Explore Configuration**: Review the `postgresql.conf` file to understand available configuration options.
3. **Set Up Development Environment**: Create a dedicated database for learning purposes and configure your preferred SQL client tools.

What's Coming Next

In our next chapter, **Cloud PostgreSQL Options,** we'll explore managed database services that eliminate infrastructure management overhead. You'll see how cloud providers like AWS RDS, Google Cloud SQL, and Azure Database build upon the same PostgreSQL foundation we've established here.

Later in **Serverless Options,** we'll discover how services like Neon and Supabase extend PostgreSQL with modern, auto-scaling architectures perfect for contemporary application development.

Key Takeaways

Universal Skills: The package manager concepts you've learned apply far beyond PostgreSQL. These skills transfer to installing development tools, system libraries, and application dependencies across your entire technology stack.

Professional Foundation: Terminal-based installation and service management are essential skills for DevOps, system administration, and production database management.

Platform Understanding: By seeing how similar tasks are accomplished across Linux, macOS, and Windows, you've gained insight into the underlying concepts that make modern database management possible.

Automation Ready: The commands and procedures you've learned can be scripted and automated, forming the foundation for infrastructure-as-code approaches and deployment automation.

The terminal installation skills you've developed today will serve you throughout your career, whether you're setting up development environments, managing production systems, or automating database deployments. In our upcoming cloud-focused chapters, you'll see how these fundamental concepts scale to enterprise environments and modern cloud architectures.

Cloud PostgreSQL Options (AWS, Azure, GCP)

Welcome to our exploration of cloud-managed PostgreSQL services. Now we'll get hands-on experience setting up PostgreSQL instances on three of the major cloud platforms: Amazon Web Services (AWS), Microsoft Azure, and Google Cloud Platform (GCP). By the end of this session, you'll have practical experience creating and connecting to managed PostgreSQL databases in the cloud.

Cloud-managed PostgreSQL services eliminate the complexity of installation, configuration, and maintenance we covered in previous chapters, while providing enterprise-grade features like automated backups, monitoring, and scaling. Let's dive into the practical steps to get PostgreSQL running on each platform.

Why Cloud PostgreSQL Services?

Before we begin, let's quickly understand what managed database services provide:

Infrastructure Management: No need to manage servers, operating systems, or PostgreSQL installations **Automated Operations**: Automated backups, patching, monitoring, and maintenance windows **Enterprise Features**: Built-in high availability, security, and compliance features **Scalability**: Easy scaling of compute and storage resources as your needs grow **Cost Efficiency**: Pay only for what you use, with no upfront hardware investments

Now let's see how to get started with each platform's PostgreSQL offering.

Amazon Web Services (AWS): RDS for PostgreSQL

AWS offers PostgreSQL through Amazon RDS (Relational Database Service), a fully managed database service that handles routine database tasks automatically. Let's create a PostgreSQL instance using the AWS Free Tier.

> AWS RDS has been merged into AWS Aurora, a proprietary, cloud-native relational database service developed by AWS that is compatible with PostgreSQL. The purpose of this book is not make any comparison between the two services, but to show how to spin up and connect PostgreSQL instances.

Creating Your AWS Account

If you don't already have an AWS account:

1. Go to aws.amazon.com
2. Click **Create an AWS Account**
3. Follow the registration process (requires a valid credit card, but we'll use free tier resources)
4. Complete identity verification
5. Choose the **Basic support plan** (free)

Creating Your First RDS PostgreSQL Instance

Step 1: Access the RDS Console

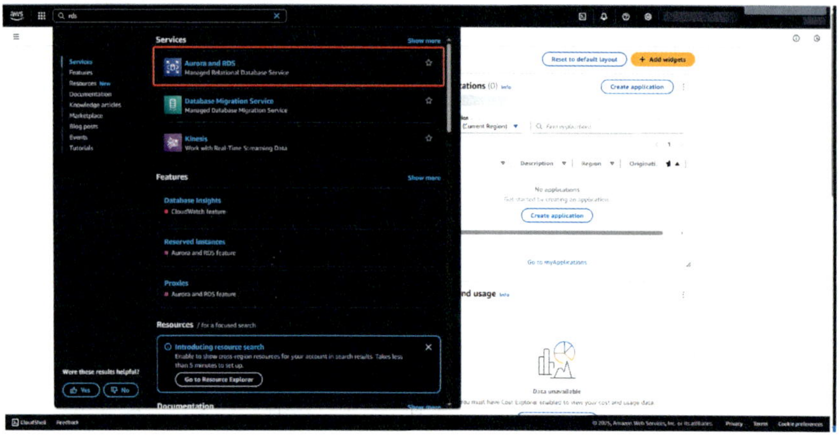

AWS console search for RDS

1. Sign in to the AWS Management Console
2. In the search bar, type **RDS** and select **Aurora and RDS**
3. Click **Create database**

Step 2: Choose Database Creation Method

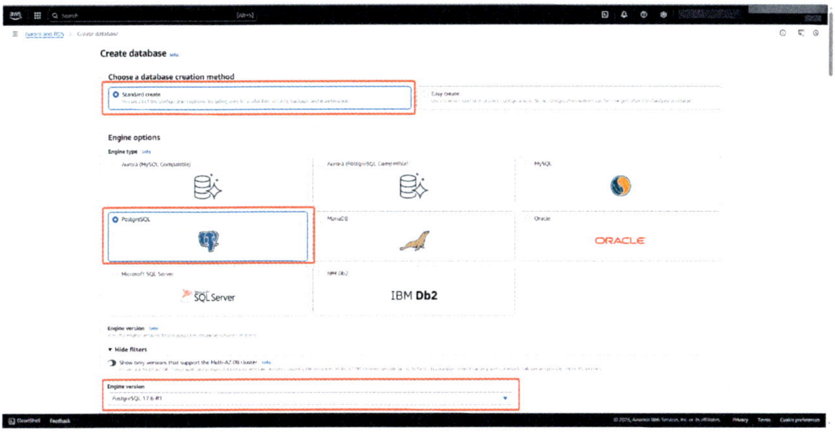

AWS console search for RDS

1. Select **Standard create** (gives us more control over configuration)
2. For **Engine type**, choose **PostgreSQL**
3. For **Engine version**, select the highest available. In my case `17.6-R1`
4. Select the **Free tier** template (this automatically configures appropriate settings)
5. Once **Free tier** is select, the only option you can choose from **Availability and durability** is **Single-AZ DB instance deployment**, which basically allows you a single instance of the database service, which is enough for our overview and tests

Step 3: Database Instance Configuration

Configure these essential settings:

DB Instance Settings: 1. **DB instance identifier**: `pg-everywhere` (or any unique name) 2. **Master username**: `postgres` (default) 3. **Credentials management**: **Self managed** to stay within the free tier (or **Managed in AWS Secrets Manager** for higher security level - and cost) 4. **Master password**: Choose a strong password and remember it. Consider using a password manager. If you select the checkbox **Auto generate password** you'll be able to see it once you're instance is created

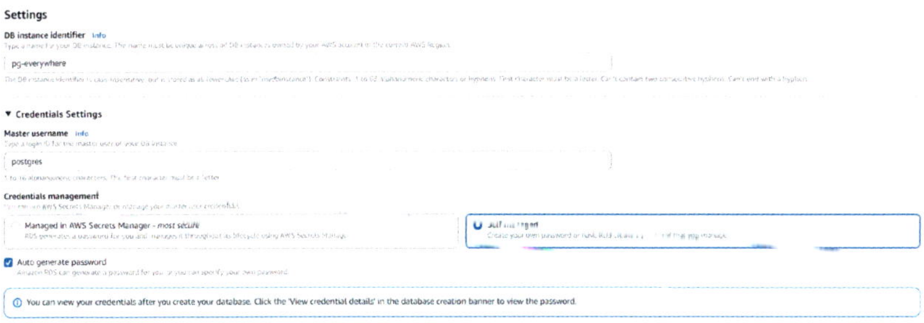

AWS DB instance settings

Instance Configuration and Storage: 1. **DB instance class**: You won't have many options within the free tier, so select `db.t4.micro` that gives you 2 vCPUs and 1GiB of RAM 2. **Storage type**: For simplicity, keep at the baseline selecting **General Purpose SSD (gp2)** 3. **Allocated storage**: 20 GiB (free tier limit)

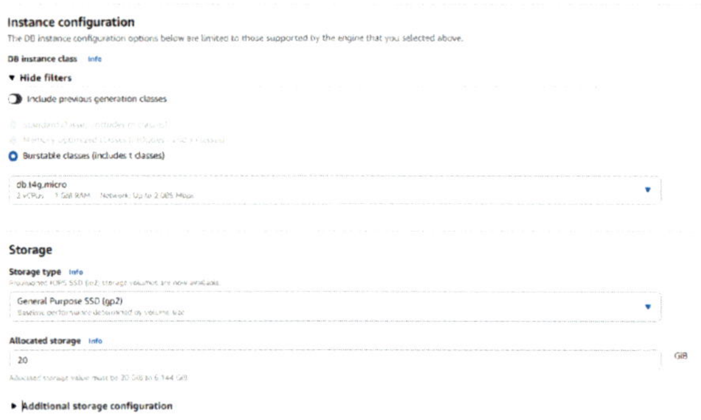

AWS DB instance settings

Step 4: Connectivity Configuration

VPC and Security: 1. **Compute resource**: You don't need to setup a connection to an EC2 instance for in this simple exercise 2. **Network type**: **IPv4** only is enough 3. **Virtual Private Cloud (VPC)**: Use default VPC 4. **DB submit group**: Also stick to the default 5. **Public access**: **Yes** to allow connection from your local machine. Interesting for our tests 6. **VPC security group**: **Create new** security group and define its name. In my case `pg-everywhere-sg` 7. **Avaliability zone**: Choose the one closer to your location. Otherwise leave it with **No preference** 8. **RDS Proxy**: For testings, you won't need a database proxy 9. **Additional configuration/Database port**: Yes, let's use 5432, why not?

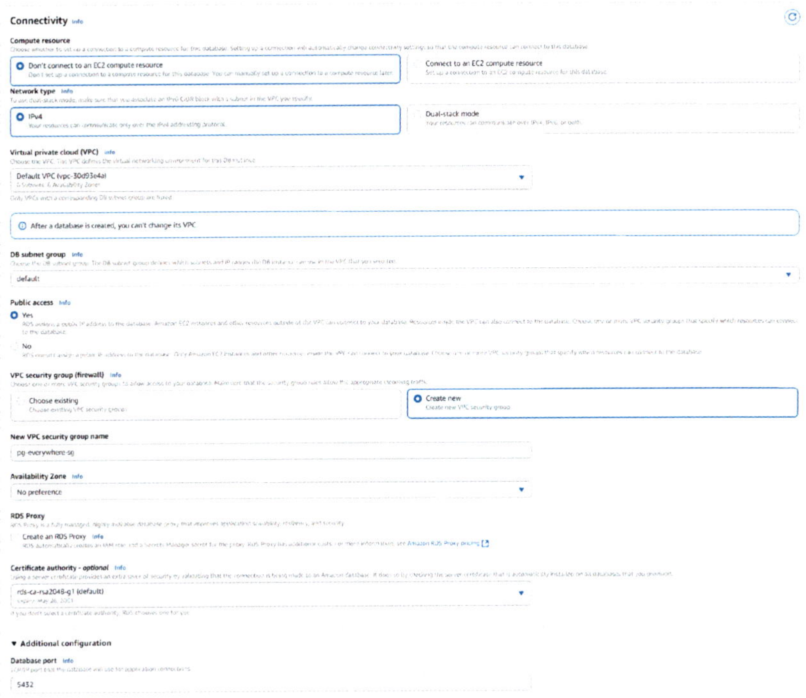

AWS DB instance settings

Step 5: Database Authentication and Monitoring 1. **Database authentication**: Password authentication 2. Leave **Performance insights** and **Enhanced monitoring** disabled for now

Step 6: Additional Configuration 1. **Initial database name**: Leave blank (we'll use the default `postgres` database) 2. **Backup retention period**: Feel free to choose from 1 to 35 days

You can either leave all other configurations as default or (recommended) spend some time getting familiar with the options.

Step 7: Create the Database

1. Review your configuration
2. Note the estimated monthly cost (should show **Free tier eligible**)
3. Click **Create database**

The creation process takes 5-10 minutes. AWS will show the status as "Creating" then "Available" when ready.

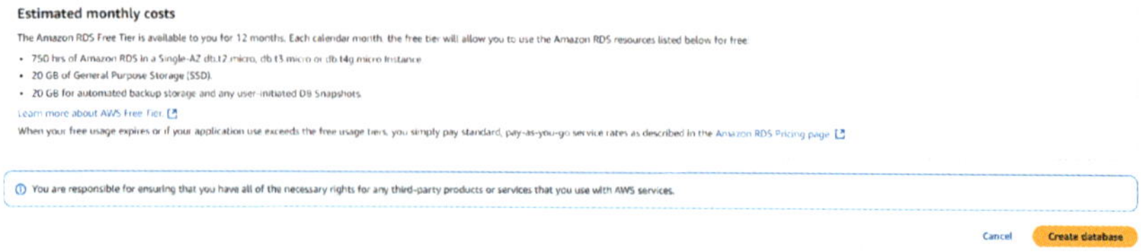

AWS DB instance settings

Copying the postgres password

While your database is being created, go to your **Databases** list page under Aurora and RDS.

There you'll see a **Creating Database** blue dialog box with a **View connection details** link. Click on it to open up a modal popup window where you can copy the auto generated password for your database user.

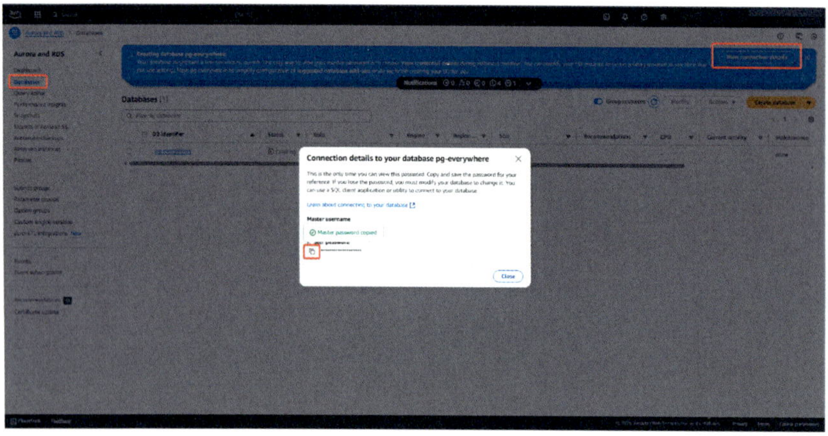

Connection details to your database

Once the database instance is up and running, there will be a green dialog box to with a similar link to open the same popup window that also contains the just created endpoint's URL. This green box is the last time you'll be able to copy your password, so make sure you don't miss it.

Configuring Security Group Access

Still while your database is being created, we need to configure network access:

Step 1: Access Security Groups

1. In the AWS Console, search for `EC2` and go to **EC2** service
2. In the left sidebar (or inside the **Resources** panel), click **Security Groups**
3. Find the security group you created in the previous process. In my case `pg-everywhere-sg` (click on the security group id link)

Step 2: Add Inbound Rule for PostgreSQL

You most like already have an Inbound rule allowing IPv4 connections of type PostgreSQL from your machine's IP only (`[YOUR-IP-ADDRESS]/32`, this 32 means all 32 bits of client's IP are checked against the given number).

In case you don't have it or intend to open to other client addresses:

1. Select your RDS security group
2. Click **Inbound rules** tab
3. Click **Edit inbound rules**
4. Click **Add rule**
5. Configure:
   - **Type**: PostgreSQL
   - **Protocol**: TCP
   - **Port**: 5432
   - **Source**: **My IP** (automatically detects your current IP address) or **Custom**, to select one you have in mind. Just please avoid the broader **Anywhere-IPv4** or **Anywhere-IPv6**, especially because we haven't configured other security layers
6. Click **Save rules**

Connecting to Your AWS RDS PostgreSQL Instance

Step 1: Get Connection Information

1. Return to the RDS console
2. Click on your database instance
3. In the **Connectivity & security** section, note:
   - **Endpoint**: (something like `pg-everywhere.abcdef123456.us-east-1.rds.amazonaws.com`)
   - **Port**: 5432

Step 2: Test Connection from Command Line

Open your terminal or command prompt on your local computer:

```
psql --host=your-rds-endpoint.region.rds.amazonaws.com --port=5432 --username=postgres --dbname=postgres
```

Replace `your-rds-endpoint.region.rds.amazonaws.com` with your actual endpoint from Step 1.

When prompted, enter the master password you set during creation.

Successful Connection Output:

```
psql (17.x, server 17.x)
SSL connection (protocol: TLSv1.3, cipher: TLS_AES_256_GCM_SHA384, compression: off)
Type "help" for help.

postgres=>
```

Step 3: Verify PostgreSQL Version

```
SELECT version();
```

This should return information about your PostgreSQL version running on AWS RDS.

Step 4: Exit the Connection

```
\q
```

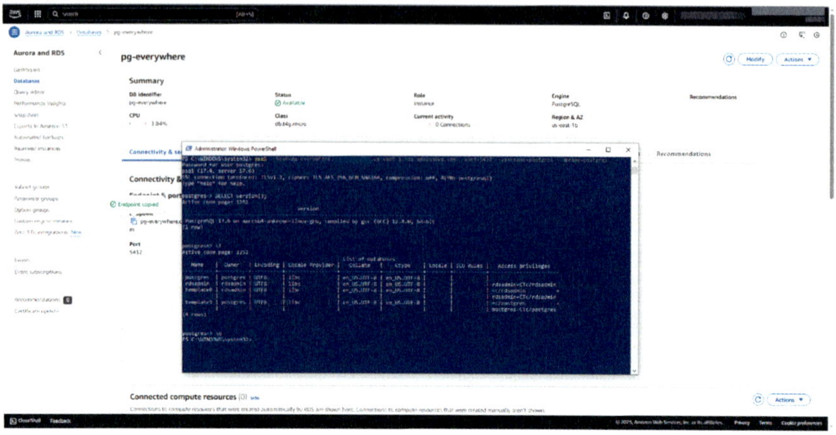

Local Powershell window successfully connected to a PostgreSQL instance on AWS

Congratulations! Your AWS RDS PostgreSQL instance is running and accessible.

Microsoft Azure: Azure Database for PostgreSQL

Azure offers PostgreSQL through Azure Database for PostgreSQL Flexible Server, providing a fully managed PostgreSQL service with built-in high availability and security.

Creating Your Azure Account

If you don't have an Azure account:

1. Go to azure.microsoft.com/free
2. Click **Try Azure for free**
3. Sign in with a Microsoft account or create one
4. Complete the registration process (requires credit card verification)
5. Get $200 in free credits plus 12 months of free services

Creating Azure Database for PostgreSQL Flexible Server

Step 1: Access the Azure Portal

1. Sign in to portal.azure.com
2. Click on the hamburger menu (top-left corner) then **Create a resource** (+ symbol in upper left)
3. Search for **Azure Database for PostgreSQL**
4. Select **Azure Database for PostgreSQL Flexible Server**

5. Click **Create**

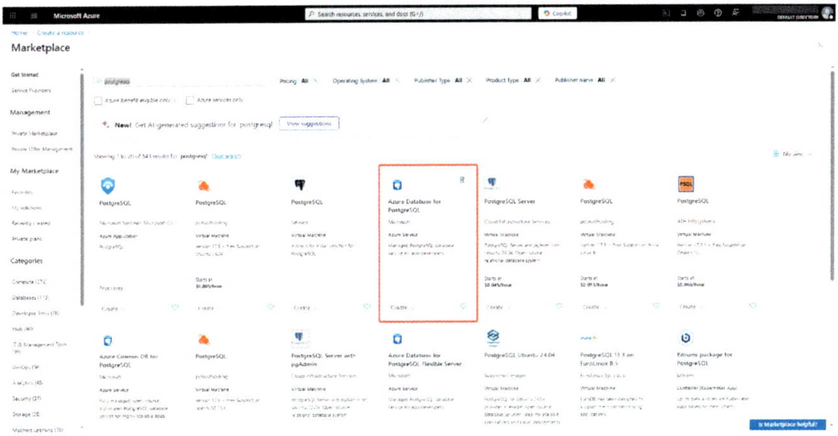

Azure Database for PostgreSQL

Step 2: Basic Configuration

Project Details: 1. **Subscription**: Select your free trial subscription 2. **Resource group**: Create new → enter `pg-everywhere-group` (or any name)

Server Details: 1. **Server name**: `pg-everywhere-server` (must be globally unique) 2. **Region**: Choose a region close to you that supports free tier 3. **PostgreSQL version**: 17 (recommended) 4. **Workload type**: **Development** (required for free tier)

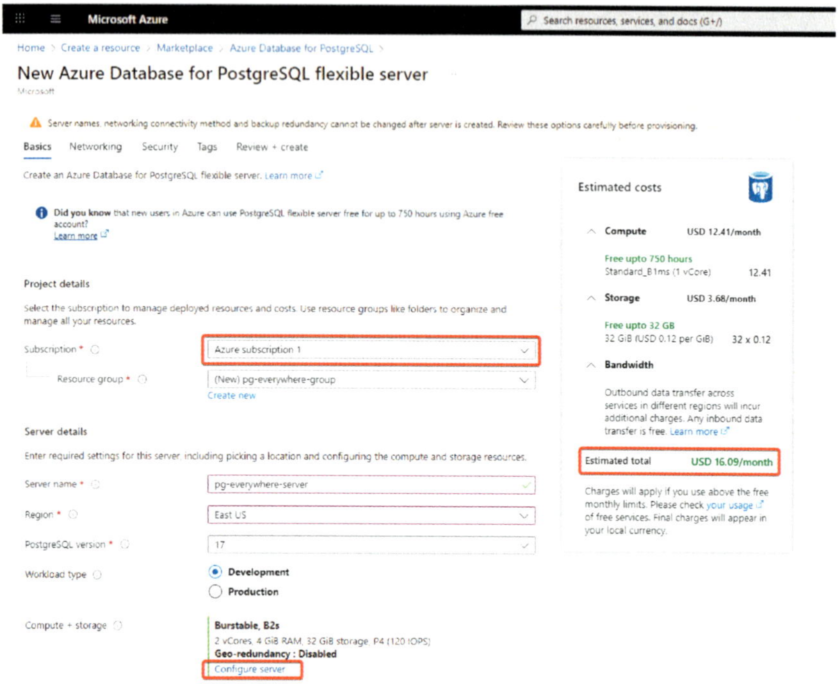

Azure PostgreSQL basic subscription configuration

It shows the estimated total cost if you use above the free monthly limits.

Step 3: Compute + Storage Configuration

1. Click **Configure server** next to **Compute + storage**
2. Ensure you select:
   - **Cluster options**: Server
   - **Compute tier**: Burstable (1-20 vCores)
   - **Compute size**: Standard_B1ms (1 vCore, 2 GiB RAM, 640 max iops)
   - **Storage type**: Premium SSD
   - **Storage size**: 32 GiB (free tier limit)
   - **Performance tier**: P4 (120 iops)
   - **Storage autogrow**: Disabled
3. Click **Save**

Step 4: Authentication

- **Authentication method**: PostgreSQL authentication only (we don't need to add extra authorisation layer for our demo, but consider it for real-world scenarios)
- **Administrator login**: `postgres` (why not?)
- **Password and Confirm password**: Enter a strong password and confirm

Step 5: Networking Configuration

1. Click **Next: Networking** or go to the next (**Networking**) tab
2. **Public access**: Enabled
3. **Connectivity method**: Public access (allowed IP addresses)
4. **Firewall rules**: Click on the link **Add current client IP address** to restrict public access to your own current IP address (change this according to your needs but for now this is enough)

At the end, you should have something like the below:

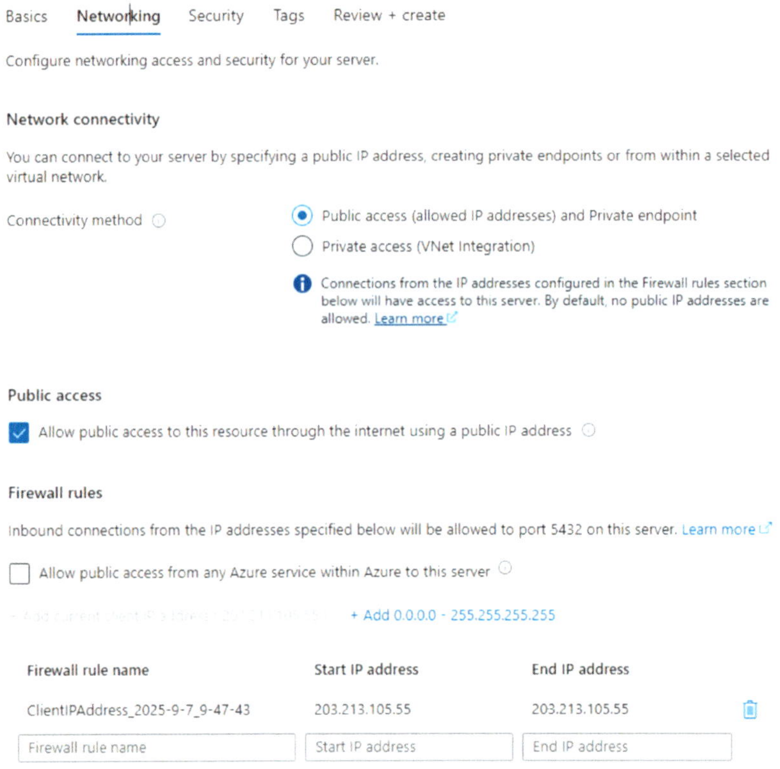

Azure simple networking configurations

Step 6: Review and Create

You don't need to change anything on **Security** and **Tags** tabs but I encourage you to take a look and decide for yourself.

1. Click **Review + create** tab
2. Verify your configuration shows **Free tier eligible**

3. Click **Create**

Deployment typically takes 3-5 minutes and you'll be redirected to the new server's page. You shouldn't expect a 100% clean and error free process. Unfortunately, general Azure quirks are beyond the scope of this book.

Connecting to Your Azure PostgreSQL Instance

Once you're lucky and deployment successfully completes, click **Go to resource**.

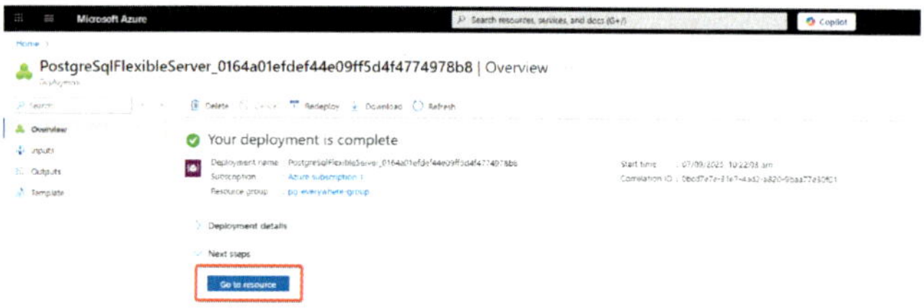

Azure resource is ready

Step 1: Get Connection Information

1. Click **Go to resource**
2. In the overview page, note:
   - **Endpoint**: (e.g., `pg-everywhere-server.postgres.database.azure.com`). Probably someone else's taken it
   - **Administrator login**: `postgres`

Step 2: Test Connection

```
psql --host=my-postgres-server-unique.postgres.database.azure.com --port=5432 --username=postgres --dbname=postgres
```

Replace the hostname with your actual server name.

Successful Connection:

```
psql (17.5 (Ubuntu 17.5-1.pgdg24.04+1))
SSL connection (protocol: TLSv1.3, cipher: TLS_AES_256_GCM_SHA384, compression: off, ALPN: postgresql)
Type "help" for help.

postgres=>
```

Step 3: Verify Installation

```
SELECT version();
```

Step 4: Exit

```
\q
```

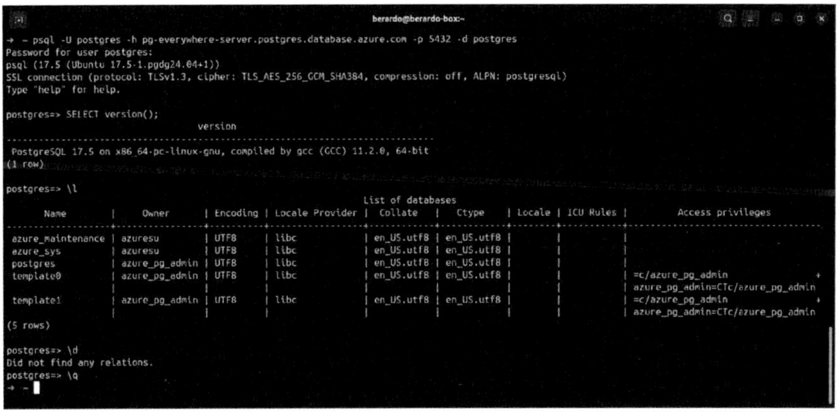

Local Ubuntu terminal successfully connected to a Ubuntu PostgreSQL instance on Microsoft Azure

Your Azure Database for PostgreSQL is now ready to use!

Google Cloud Platform (GCP): Cloud SQL for PostgreSQL

Google Cloud offers PostgreSQL through Cloud SQL, a fully managed database service. Note that Cloud SQL doesn't have a traditional "free tier" like AWS and Azure, but Google Cloud provides $300 in free credits for new accounts that can cover several months of small database usage.

Creating Your Google Cloud Account

If you don't have a Google Cloud account:

1. Go to cloud.google.com
2. Click **Get started for free**
3. Sign in with a Google account
4. Select your **Country** and click **Agree & continue**
5. Accept terms and complete verification (requires credit card)
6. Receive $300 in free credits (valid for 90 days)

Running on your Google account on Google Chrome, you should see your profile loaded and possibly even your default payment method. In this case, just hit the **Start free** button.

Creating a Cloud SQL PostgreSQL Instance

Step 1: Access Cloud SQL

1. Open the Google Cloud Console
2. Search for `SQL` in the search bar
3. Click **SQL Product** from the results
4. Click **Create Instance**. Alternatively click on Start with Sandbox in the first and minimal option of the PostgreSQL Get started tab.

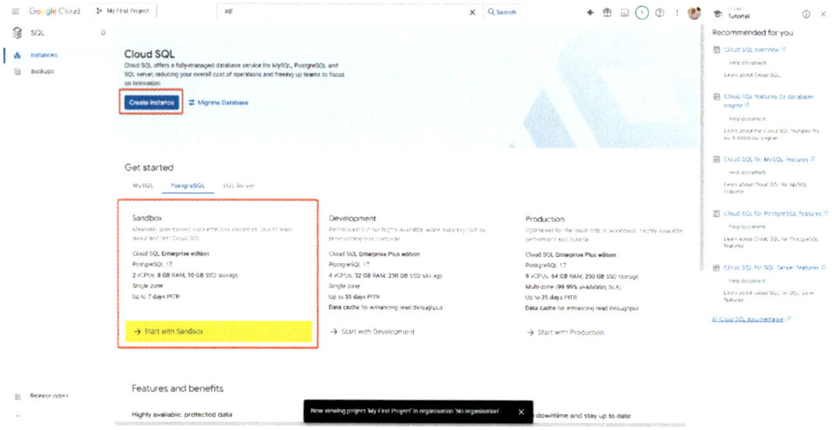

Google Cloud SQL initial page

Step 2: Choose Database Engine

1. Click **Choose PostgreSQL**
2. If asked, click **Enable API**

Step 3: Choose a Cloud SQL edition

Instance Info: 1. Select **Enterprise** (99.95% availability SLA), the cheapest option 2. **Edition preset**: Sandbox 3. **Instance version**: PostgreSQL 17 4. **Instance ID**: `pg-everywhere` 5. **Password**: Set a strong password for the `postgres` user 6. **Region**: Choose a region close to your location 7. **Zonal availability**: Single zone (more cost-effective)

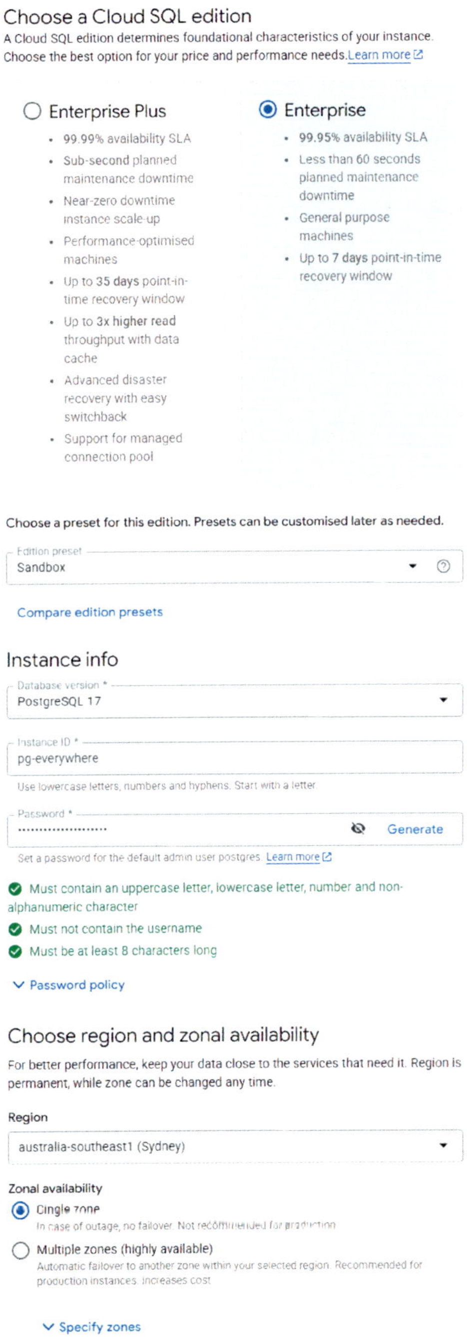

GCP Choose a Cloud SQL edition configuration

Step 4: Customise your instance

Click **Show configuration options** to customise:

Machine configuration: 1. **Machine**: **Shared core**. Not ideal in production environments, but we're just testing, so better go with the cheapest option 2. **vCPUs**: 1 vCPU, 0.614 GB 3. **Storage type**: SSD 4. **Storage capacity**: 10 GB

This configuration provides the smallest possible instance for cost efficiency.

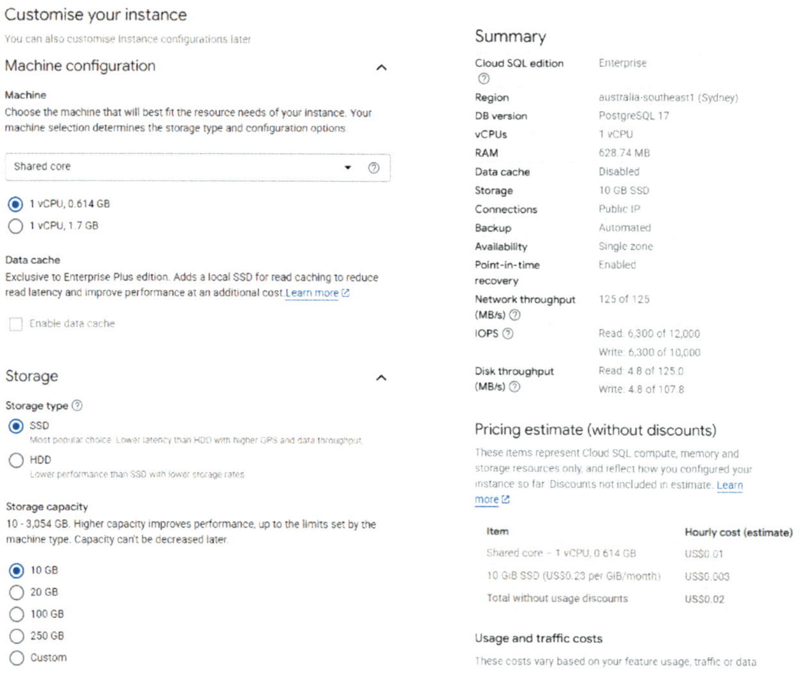

Cloud SQL minimal machine configuration

Step 5: Connections

1. **Public IP**: Enable
2. **Authorized networks**: Click **Add network**

- **Name**: `My Computer`
- **IP range**: Click **Use my IP** to load it in the text field then add `/32` to have the most specific CIDR possible
- Click **Done**

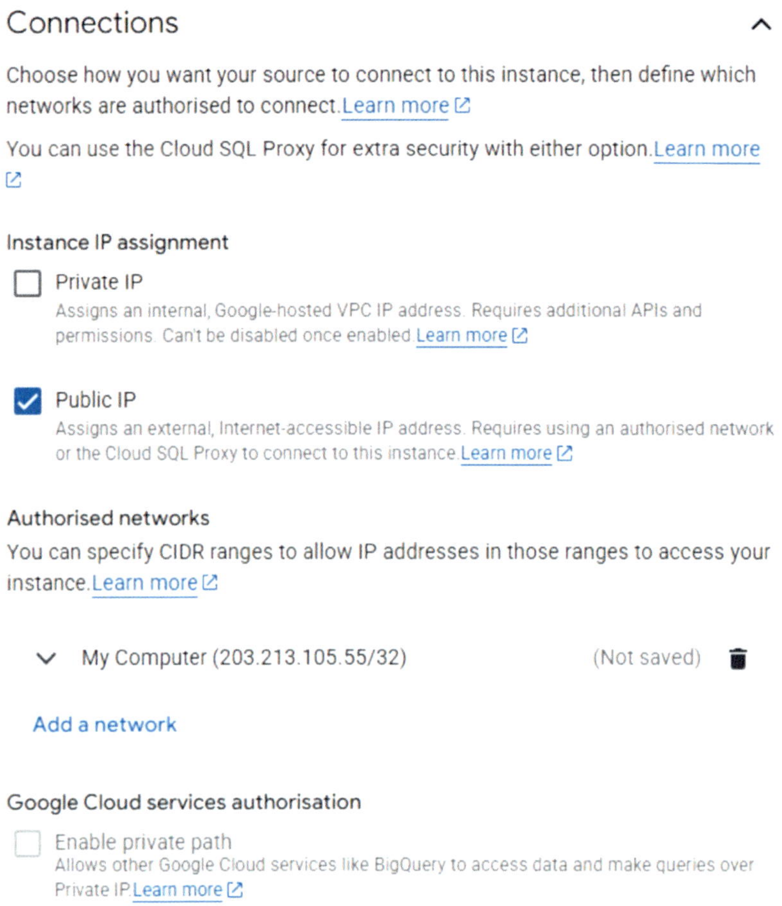

Cloud SQL minimal machine configuration

Step 6: Security

You can leave the default **Server certificate authority mode: Google-managed internal certificate authority**.

Step 7: Data Protection

- **Automated backups**: Enable. Leave default **Number of days** and **Backup window**
- **Point-in-time recovery**: Enable. Leave default **Days of logs**

Step 8: Check other configurations and Create instance

1. Review configuration and estimated monthly cost

2. Click **Create Instance** button

Instance creation takes 5-10 minutes.

Connecting to Your Cloud SQL Instance

Step 1: Get Connection Information

1. Once the instance is created, click on the instance name
2. In the **Connect to this instance** section, note:
    - **Public IP address**
    - **Connection name** (for Cloud SQL Proxy)

Step 2: Connect Using psql

```
psql --host=YOUR_INSTANCE_PUBLIC_IP --port=5432 --username=postgres --dbname=postgres
```

Replace `YOUR_INSTANCE_PUBLIC_IP` with the actual IP address from the Cloud Console.

Alternative: Using Cloud Shell

Google Cloud provides a browser-based terminal:

1. In the Cloud Console, click the **Cloud Shell** icon (terminal symbol)
2. Once Cloud Shell loads, use the `gcloud sql connect` command:

```
gcloud sql connect my-postgres-instance --user=postgres
```

Successful Connection:

```
psql (17.4, server 17.5)
SSL connection (protocol: TLSv1.3, cipher: TLS_AES_256_GCM_SHA384, compression: off, ALPN: postgresql)
Type "help" for help.

postgres=>
```

Step 3: Verify Installation

```
postgres=> SELECT version();
Active code page: 1252
                                          version
------------------------------------------------------------------------------------------
 PostgreSQL 17.5 on x86_64-pc-linux-gnu, compiled by Debian clang version 12.0.1, 64-bit
(1 row)
```

Step 4: Exit

```
\q
```

Bonus: Connecting to Your Cloud SQL Instance through pgAdmin 4

Open pgAdmin 4 on any operating system you have it installed.

Step 1: Register servers under new Server Group 1. Right click **Servers** then click **Create / Server groups** 2. Type your server group name: `cloud-providers` 3. Right click cloud-providers then click **Register / Server** 4. On **General** tab, type a **Name** for your cloud provider: `gcp-server`

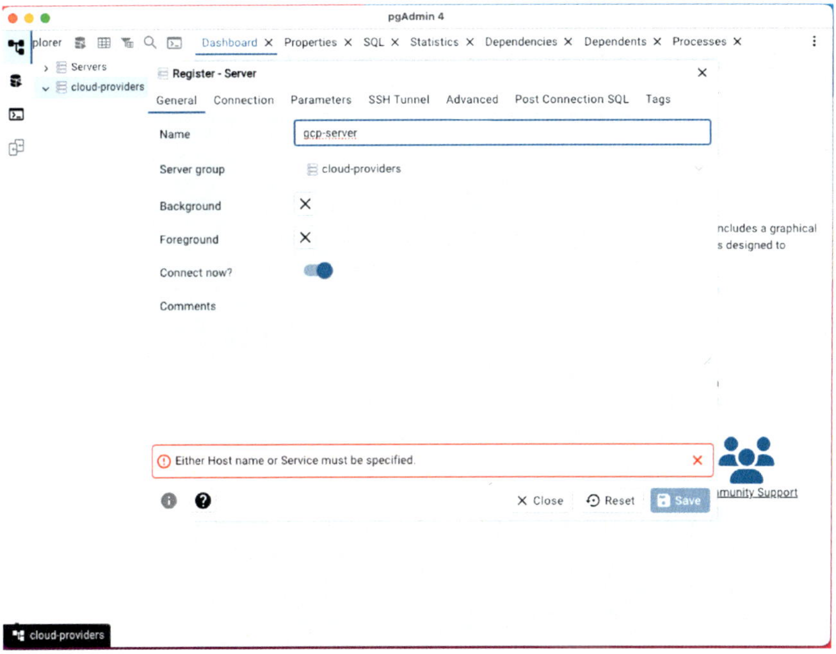

New server group and server name on pgAdmin 4

Step 2: Connect external servers 1. Click **Connection** tab 2. Type the same IP address on last psql connection `YOUR_INSTANCE_PUBLIC_IP` 3. Type your defined postgres' password 4. Click on **Save**

Your Google Cloud SQL PostgreSQL instance is now operational!

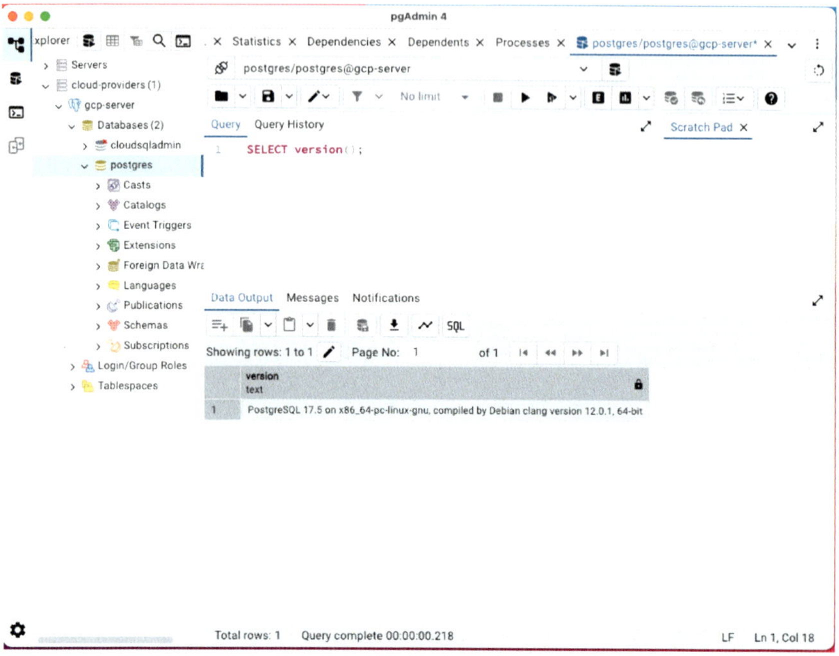

New server group and server name on pgAdmin 4

Important Cost Management

Since you're using cloud services, here are essential cost management tips:

AWS RDS

- **Monitor usage**: Stay within 750 hours/month free tier limit
- **Stop/Start instances**: You can stop instances for up to 7 days to save hours
- **Delete when done**: Don't forget to delete test instances

Azure Database

- **12-month limit**: Free tier is valid for 12 months from account creation
- **Resource monitoring**: Monitor usage in Azure portal
- **Auto-shutdown**: Consider setting up auto-shutdown for development instances

Google Cloud SQL

- **Credit monitoring**: Keep track of your $300 free credits
- **Smallest instance**: Use the smallest possible instance size for learning
- **Delete unused instances**: Always clean up instances you're not using

Cleanup Instructions

AWS RDS: 1. Go to RDS Console → Databases 2. Select your instance → Actions → Delete 3. Uncheck "Create final snapshot" and "Retain automated backups" 4. Type "delete me" and confirm

Azure Database: 1. Go to your resource group 2. Select the PostgreSQL server 3. Click "Delete" and confirm

Google Cloud SQL: 1. Go to Cloud SQL → Instances 2. Select your instance → Delete 3. Type the instance ID to confirm

Next Steps

Congratulations! You now have hands-on experience with all three major cloud PostgreSQL offerings. You've learned how to:

- Create managed PostgreSQL instances on AWS RDS, Azure Database, and Google Cloud SQL
- Configure network access and security settings
- Connect to your cloud databases from your local machine
- Perform basic database operations

What's Coming Next

In our next chapter, **Serverless Options,** we'll explore modern PostgreSQL-as-a-Service platforms like Neon, Supabase, and TigerData that take cloud databases even further with serverless architectures, automatic scaling, and developer-friendly features.

Following that, in **Container Fundamentals,** we'll see how PostgreSQL can be deployed using Docker and Kubernetes for maximum flexibility and portability.

Key Takeaways

Cloud Advantages: Managed services eliminate infrastructure complexity while providing enterprise-grade features out of the box.

Platform Similarities: Despite different interfaces, all three platforms provide similar core functionality for managed PostgreSQL.

Cost Awareness: Free tiers and credits make learning affordable, but always monitor usage and clean up resources.

Production Ready: These managed services are production-grade solutions used by companies worldwide for mission-critical applications.

Serverless Options (Neon, Supabase, TigerData)

Welcome to our exploration of the cutting edge of PostgreSQL-as-a-Service: serverless database platforms. Now we'll discover how services like Neon, Supabase, and TigerData represent the next evolution in database services, offering capabilities that go beyond traditional managed databases.

These platforms embody truly modern approaches to database management - from database branching like Git repositories, to real-time APIs, to specialized time-series optimizations. Let's dive into what makes serverless PostgreSQL special and get hands-on experience with each platform's unique offerings.

> Disclaimer: There are countless PostgreSQL offerings out there. The three serverless options by no means should be treated as an attempt of build a comprehensive list or an intended classification of the "best" ones. They are purely the three options I know and thought that'd be worth sharing.

Understanding Serverless PostgreSQL Architecture

Before we create accounts and databases, let's understand what distinguishes serverless PostgreSQL from the cloud managed services we explored in the previous chapter.

Traditional vs. Serverless Architecture

Traditional Cloud Databases (AWS RDS, Azure, GCP): - **Fixed Instance Model**: You provision specific compute and storage resources - **Always-On**: Resources run continuously, whether you're using them or not - **Manual Scaling**: Requires intervention to increase or decrease capacity - **Standard Features**: Focus on reliability, backups, and traditional database operations

Serverless Databases: - **Separation of Compute and Storage**: Resources scale independently - **Auto-Scaling**: Compute resources scale to zero when idle and instantly scale up under load - **Developer-Centric Features**: Built-in APIs, real-time capabilities, developer tools - **Modern Workflows**: Features like database branching, instant provisioning, and integrated development environments

Key Advantages of Serverless PostgreSQL

Cost Efficiency: Pay only for actual compute time used, not idle resources **Instant Provisioning**: New databases and branches created in seconds **Modern Development Patterns**: Features designed for modern application development **Zero Maintenance**: No server management, patching, or infrastructure concerns **Advanced Features**: Capabilities not available in traditional databases

Now let's see how each platform implements these concepts differently.

Neon: Database Branching and True Serverless

Neon represents the most PostgreSQL-native serverless experience, introducing the revolutionary concept of database branching - treating your database like a Git repository with branches for different environments and use cases.

What Makes Neon Special

Database Branching: Create instant copies of your database for testing, feature development, and staging environments.

True Serverless Architecture: Compute and storage are completely separated, allowing instant scaling from zero to full capacity.

PostgreSQL Compatibility: 100% PostgreSQL wire protocol compatibility with support for PostgreSQL versions 14, 15, 16, and 17.

Generous Free Tier: 10 projects, 10 branches per project, 0.5GB storage, and 191.9 compute hours per month.

Creating Your Neon Account and First Database

Step 1: Account Registration

1. Visit neon.com
2. Click **"Sign up"**
3. Choose to sign up with **Email**, **GitHub**, **Google**, or **Microsoft** account
4. Complete the verification process

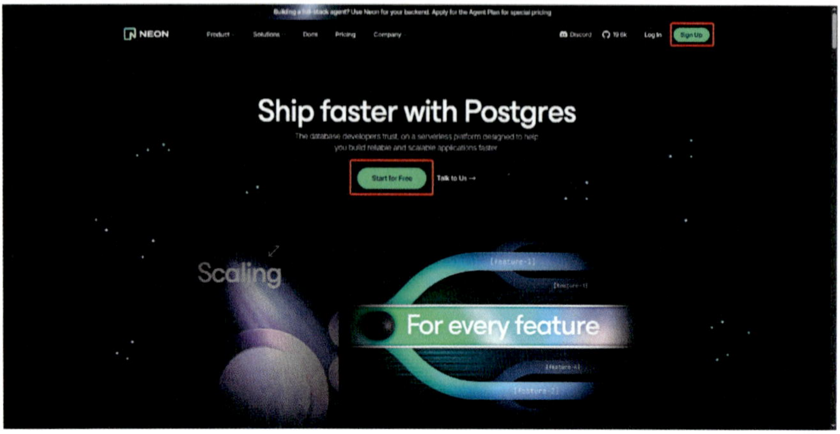

Neon homepage

Step 2: Create Your First Project

After signing up, you'll be directed to the project creation screen:

1. **Project Name**: Enter `pg-everywhere` (or any descriptive name)
2. **PostgreSQL Version**: Select **17**
3. **Cloud service provider**: AWS or Azure (yes, serverless options run on major cloud providers)
4. **Region**: Choose the region closest to your location for optimal performance
5. Click **"Create Project"**

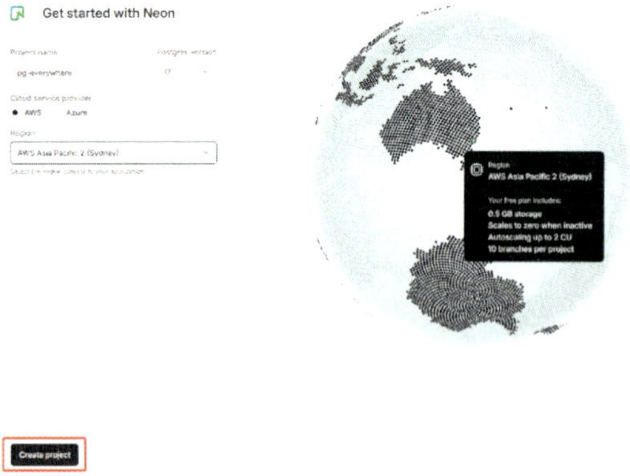

Create new Neon project

Project creation takes just a few seconds - this instant provisioning is a key serverless advantage!

Step 3: Explore the Neon Dashboard

Once your project is created, you'll see the Neon dashboard featuring:

- **Connect to your database**: Ready-to-use connection strings for several different clients (including **psql** and **Parameters only**, useful for easy copy-paste on tools like pgAdmin 4)
- **Branches**: It starts with two branches, **production** and **development** that you can use to fork your project's database structure and actual data. You can have up to 10 branches in a free plan
- **Data API (beta)**: To create an instant REST API for your database
- **Monitoring**: Real-time compute and storage usage
- **SQL Editor**: Built-in query interface

Connecting to Your Neon Database

Step 1: Get Connection Information

1. In your project dashboard, click on **connect** button either on top right corner or in the **Connect to your database** box.
2. **Branch**: Select **development** branch
3. **Database**. Select **Create new** and:
   - **Database name**: for example `pg-everywhere-db`
   - **Owner**: leave the default **neondb_owner**
   - Click **Create**

Step 2: Connect From psql Using Connection String

At this stage, I'm pretty sure you know how to connect your local psql to any external PostgreSQL server. If not, review the last chapter to train how to do connect to AWS, Azure or GCP.

Interestingly, psql has two more connection options. One is using a connection string. One is via connection string and the other is a passwordless auth.

Let's take the opportunity that Neon is showing us how to use them and learn these two interesting options.

If you have ever written code in languages like Java using JDBC, you should be familiar with this *URI-style* syntax. Something like:

```
protocol://username:password@server.domain/resource/path?parameter=value
```

Using psql, you can set **postgresql** as your protocol and fit the rest as follows:

```
psql 'postgresql://neondb_owner:[YOUR-PASSWORD]@ep-something.your-region.aws.neon.tech/pg-everywhere-db?sslmode=require&channel_binding=require'
```

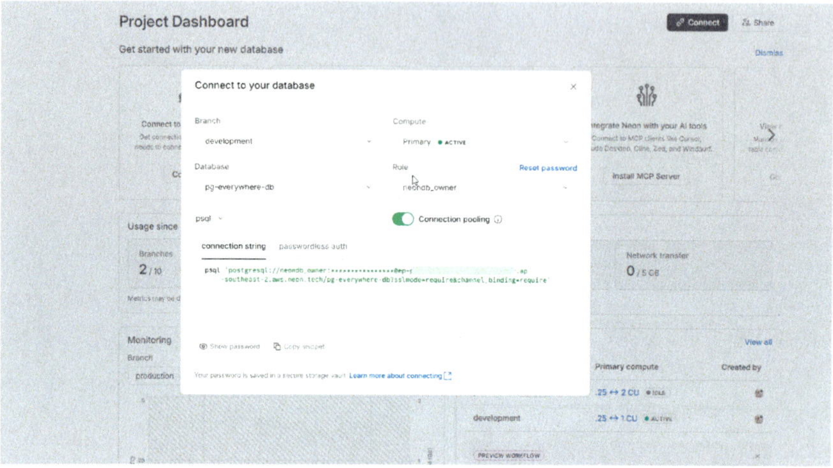

How to grab the connection string on Neon

Click on **Copy snippet** and use it to connect.

Step 3: Connect From psql Using Passwordless auth

Let's try the other method: passwordless auth.

Coming back to Neon's **Connect to your database** dialog box, if click on **passwordless auth** tab you'll see just:

```
psql -h pg.neon.tech
```

This super simple connection explores your browser session to connect. After running to code above, you'll see a message like:

```
NOTICE:  Welcome to Neon!
Authenticate by visiting (will expire in 2m):
    https://console.neon.tech/psql_session/[A RANDOM HEX NUMBER]
```

Go ahead and open this link on the same browser session you are logged into Neon. You'll first be presented a screen like this:

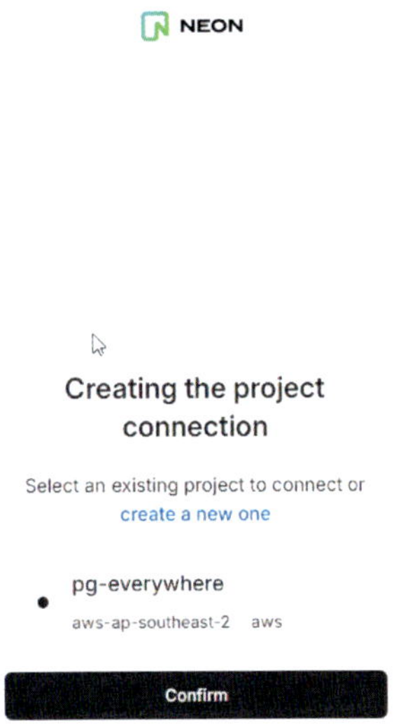

Create a passwordless connection to Neon PostgreSQL

Select your database and click **Confirm**. Then you'll see the branch selection screen:

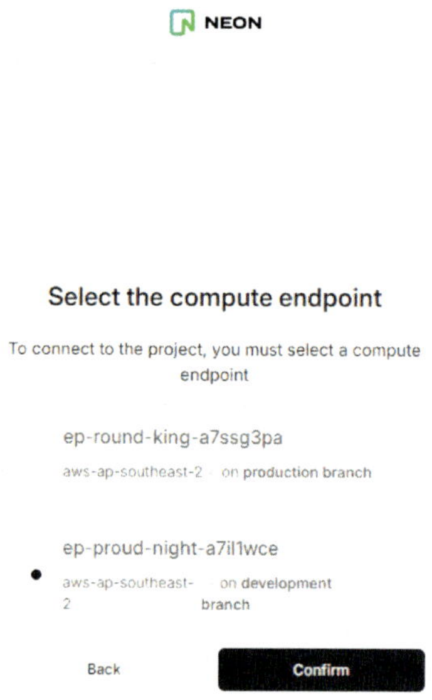

Select your database "branch"

Select your desired branch and **Confirm** again.

The final screen is an indication that your psql is now connected.

You are now passwordless connected to Neon PostgreSQL

Come back to psql and see the magic:

```
NOTICE:  Connected to database, session_id: d5583ccf-01bb-4026-a9b4-03a7a296e823
psql (17.4, server 17.5 (1b53132))
SSL connection (protocol: TLSv1.3, cipher: TLS_AES_256_GCM_SHA384, compression: off, ALPN: postgresql)
Type "help" for help.

joseb=>
```

Seriously, what can be easier?

Step 4: Test Serverless Features

Let's verify Neon's unique capabilities:

```sql
-- Check PostgreSQL version
SELECT version();

-- Create a simple table to test branching later
CREATE TABLE serverless_test (
    id INTEGER PRIMARY KEY GENERATED ALWAYS AS IDENTITY,
    platform VARCHAR(50),
    feature VARCHAR(100),
    created_at TIMESTAMP DEFAULT CURRENT_TIMESTAMP
```

```
);

-- Insert test data
INSERT INTO serverless_test (platform, feature)
VALUES ('Neon', 'Database Branching'), ('Neon', 'Serverless Compute');

-- Verify data
SELECT * FROM serverless_test;
```

Step 4: Experience Database Branching

This is Neon's killer feature - let's create a branch:

1. In the Neon dashboard, click **"Create Branch"**
2. **Branch Name**: `branching-test`
3. **Branch From**: Select your main branch
4. Click **"Create Branch"**

You now have an instant copy of your entire database! The branch includes all data, schema, and can be used independently for testing or development.

Supabase: The Backend-as-a-Service PostgreSQL Platform

Supabase positions itself as the "open source Firebase alternative," providing not just a PostgreSQL database but a complete backend-as-a-service platform with real-time capabilities, authentication, and auto-generated APIs.

What Makes Supabase Unique

Complete Backend Platform: PostgreSQL database plus authentication, real-time subscriptions, storage, and auto-generated APIs.

Real-Time Everything: Built-in real-time subscriptions for database changes, perfect for collaborative applications.

Auto-Generated APIs: RESTful and GraphQL APIs automatically generated from your database schema.

Generous Free Tier: 50,000 monthly active users, 500MB database storage, 5GB bandwidth, and 1GB file storage.

Developer Experience: Exceptional tooling including built-in SQL editor, table editor, and API documentation.

Creating Your Supabase Account and Project

Step 1: Account Setup

1. Go to supabase.com
2. Click **"Start your project"**

3. **Sign up with GitHub** (recommended for best integration)
4. Authorize Supabase to access your GitHub account

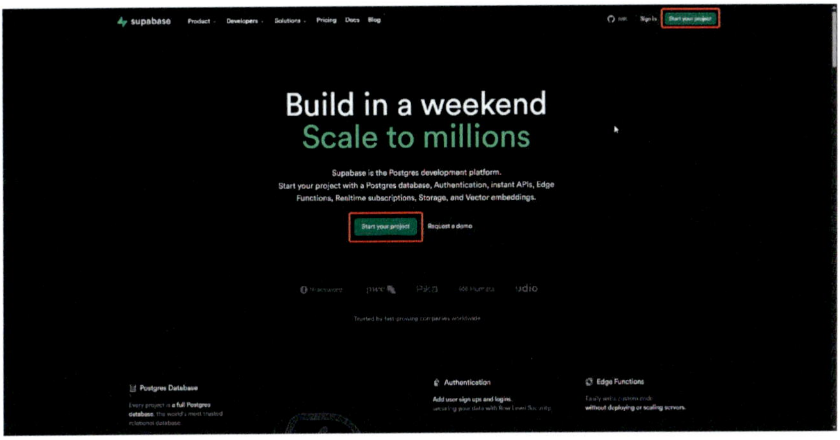

Supabase homepage

Step 2: Create Your Organisation

1. Click **"New Project"** (or if this is your first project, the creation wizard starts automatically)
2. **Name**: Create an organisation name
3. **Type**: Define what best describe your organisation
4. **Plan**: Select your plan. Recommended: **Free = $0/month**
5. Click **Create organization**

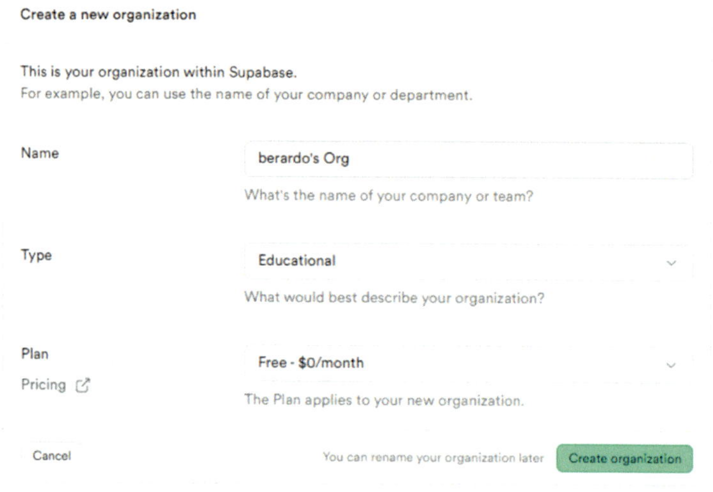

New organisation on Supabase

Step 3: Create New Project

1. **Project Name**: `Postgres Everywhere` or any name you want
2. **Database Password**: Click **"Generate a password"** and **save this password securely**
3. **Region**: Choose the region closest to you
4. Under **Security Options**:
    - **What connections do you plan to use?**
    - As you saw in previous cloud offerings, you should be good with selecting **Only Connection String**. However, I recommend you to try the **Data API + Connection String** to see Supabase automatic REST API generation, which is super cool
    - **Data API configuration**
    - Leave it at **Use public schema for Data API** for our tests and your initial development, but you can opt for dedicated API schema where you define which tables are visible through the REST API
5. Under **Advanced Configuration**:
    - **Postgres Type**: Leave it as Postgres
6. Click **"Create new project"**

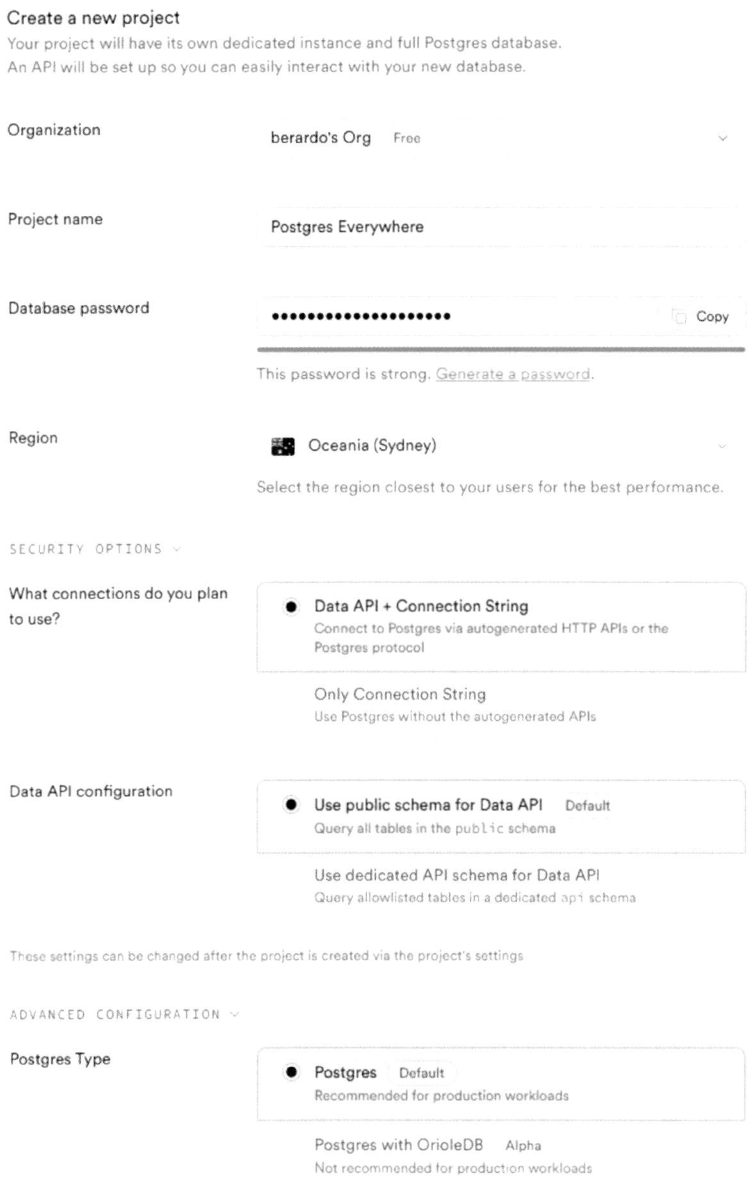

New organisation on Supabase

Supabase should be very swift provisioning your PostgreSQL instance and setting up all backend services.

Step 3: Explore the Supabase Dashboard

Your Supabase dashboard provides access to:

- **Table Editor**: Visual database management
- **SQL Editor**: Advanced query interface
- **Authentication**: User management and auth configuration

- **Storage**: File upload and management
- **API**: Auto-generated documentation for your database

You'll notice some overlap with Neon offering. As they compete, they start to feature similar capabilities. For example, Supabase also invested in the concept of database "branches".

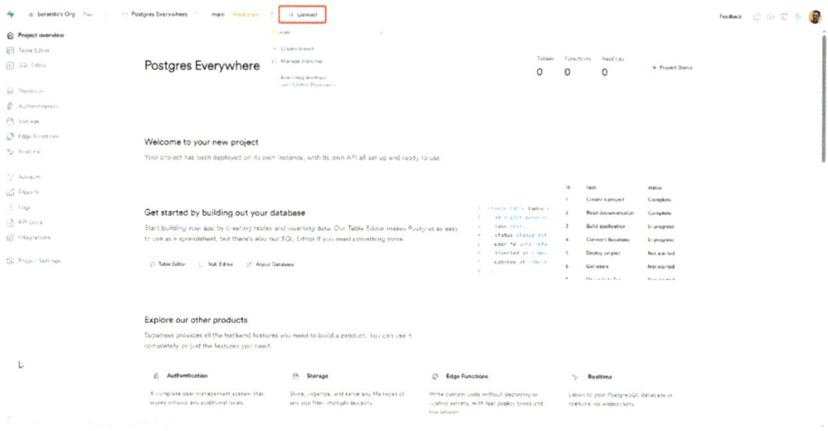

Supabase project dashboard

Connecting to Your Supabase Database

Step 1: Get Connection Information

1. In your project dashboard, click **Connect** on the top bar (as per previous image)
2. On **Connection String** tab
   - Set **Type** to psql (but you already know how to use the URI).
   - The **Direct connection** box has similar instructions to what we've been doing since the terminal installations. You can expand **View parameters** to see each parameter individually. This is good for human-driven connections where an actual person is requesting a long-lived connection to operate in a database
   - Because Supabase is application-driven, it also has a configuration using **Transaction pooler**, for short connection (usually from serverless functions)
   - There's also this **Session pooler**, which uses a connection pool, like the transaction pooler, but keeps your entire session against the same server connection. So prefer this little box.

- Click on copy icon next to **Session pooler**.

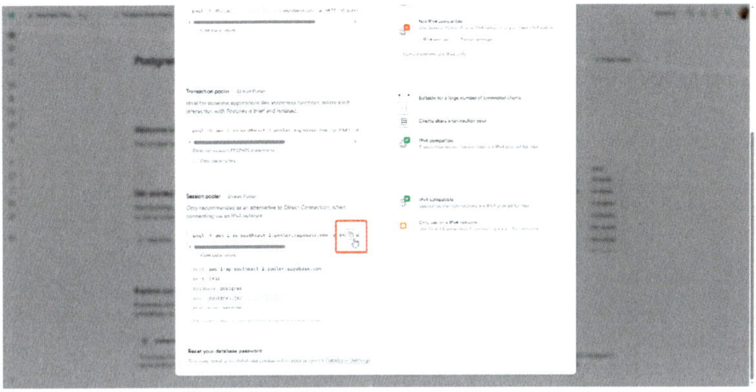

Supabase Session pooler details

Step 2: Connect Using psql

```
psql -h aws-1-ap-southeast-2.pooler.supabase.com -p 5432 -d postgres -U postgres.[your-user-number]
```

When asked, provide the password you set when created the database.

Step 3: Test Real-Time Capabilities

Let's create a table that showcases Supabase's real-time features:

```
-- Create a real-time enabled table
CREATE TABLE real_time_messages (
    id SERIAL PRIMARY KEY,
    message TEXT NOT NULL,
    user_name VARCHAR(50),
    created_at TIMESTAMP DEFAULT CURRENT_TIMESTAMP
);

-- Insert test data
INSERT INTO real_time_messages (message, user_name)
VALUES
    ('Hello from Supabase!', 'system'),
    ('Real-time PostgreSQL is awesome', 'developer');

-- Check the data
SELECT * FROM real_time_messages ORDER BY created_at DESC;
```

Step 4: Explore Auto-Generated APIs

1. In your Supabase dashboard, click **API Docs** in the left sidebar

2. Go to **Tables and Views** in the API documentation
3. Click on your `real_time_messages` table
4. Explore the auto-generated REST endpoints and the Supabase Javascript library they use to explain what your (Javascript) app could do with it.

This auto-API generation is what makes Supabase incredibly powerful for rapid application development.

TigerData: The Evolution of TimescaleDB for Modern PostgreSQL

TigerData (formerly TimescaleDB/Timescale) represents a significant evolution beyond traditional time-series databases. In June 2025, the company rebranded from Timescale to TigerData to reflect their transformation from a specialized time-series platform into what they call "the modern PostgreSQL for transactional, analytical, and agentic workloads."

What Makes TigerData Unique

Evolution Beyond Time-Series: While maintaining its time-series optimization roots, TigerData now positions itself as a comprehensive PostgreSQL platform supporting AI, analytics, and traditional transactional workloads.

Tiger Cloud Platform: The managed cloud service (formerly Timescale Cloud) offers separation of compute and storage, auto-scaling, and integration with data lakes and modern analytics architectures.

AI and Vector Capabilities: Built-in support for vector search with pgvector and pgvectorscale extensions, plus pgai for embedding generation and LLM operations directly in PostgreSQL.

Generous Free Tier: 30-day trial with $300 credits, though long-term use requires paid plans for production workloads.

TigerData's Modern Architecture

Hypertables and Time-Series Optimization: - Automatic time-based partitioning that appears as regular PostgreSQL tables - Hybrid row-columnar storage for optimal query performance - Continuous aggregates that automatically refresh materialized views - Data compression and tiering to S3 for cost optimization

Tiger Lake Integration: - Seamless integration with data lakes using Apache Iceberg format - Real-time data streaming to S3 with zero-copy operations - Query data across PostgreSQL and lake storage as a unified system

Getting Started with TigerData

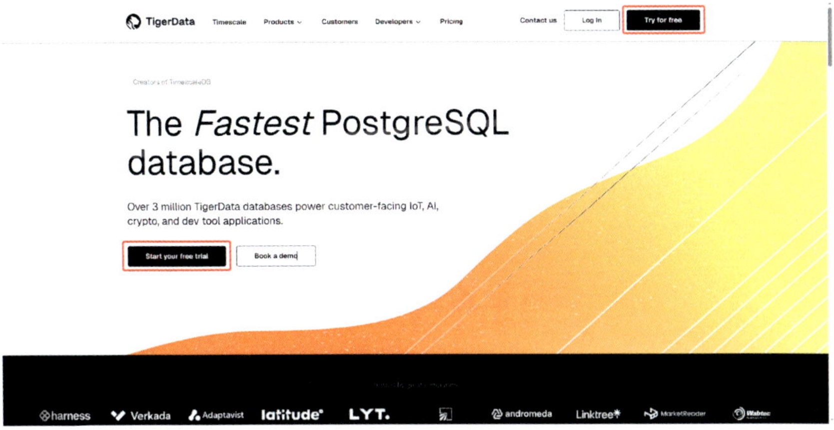

TigerData homepage

Step 1: Account Setup 1. Visit tigerdata.com 2. Click **Try for free** or **Start your free trial** 3. On the next page, fill up the form or click **Sign up with Google** - TigerData runs by default on AWS, but you can choose Azure or GCP on this page

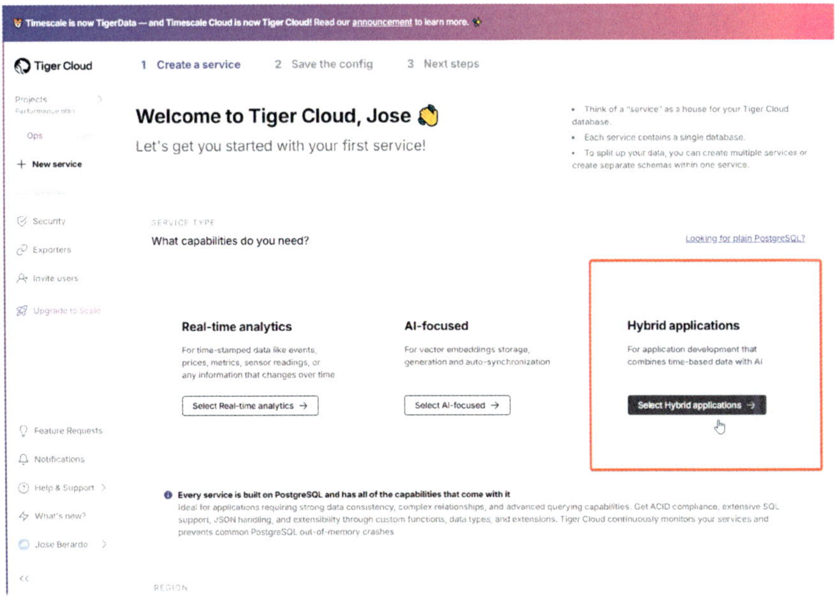

TigerData initial logged in page

Step 2: Create a Service

Once you're signed in, you'll land on the **Create a service** page. Let's explore their offering while we can enjoy the 30-day free allowance.

1. Click **Select Hybrid applications**
2. **Region**: Set the region close to you and hit **Save and continue**
3. **Compute size**: Choose the smallest options as we don't need to go fancy then hit **Save and continue**. You'll see something like from `$0.06800` to `$0/hour` during free trial (remember to drop your instance before the 30 day - or be a happy payer)
4. **Environment and Availability**: Keep the default **Development** the recommended **No replica**
5. **Power-Ups**: You won't need them. Just hit **Save and continue**
6. Click **Create service**

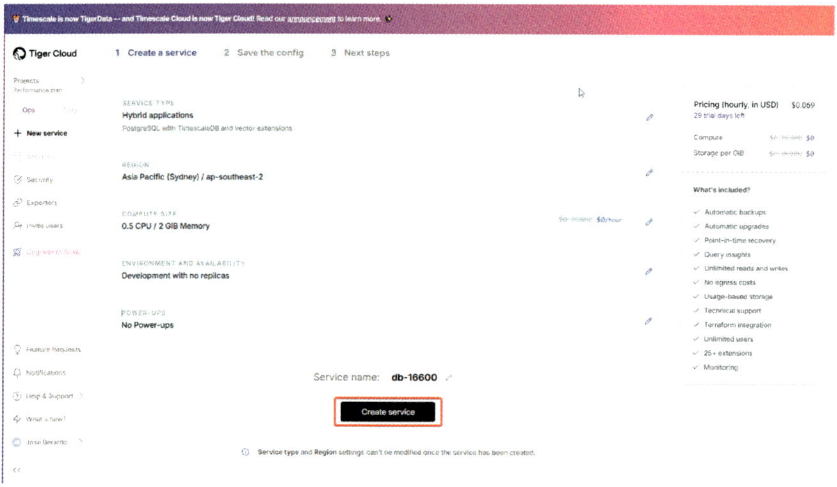

Create a service summary on Tiger Cloud

Step 3: Save the config

1. Click **Download the config** in the **Download your database config** box
2. You can also copy the Connection string right below

Notice that, for the first time, 5432 is not our default port.

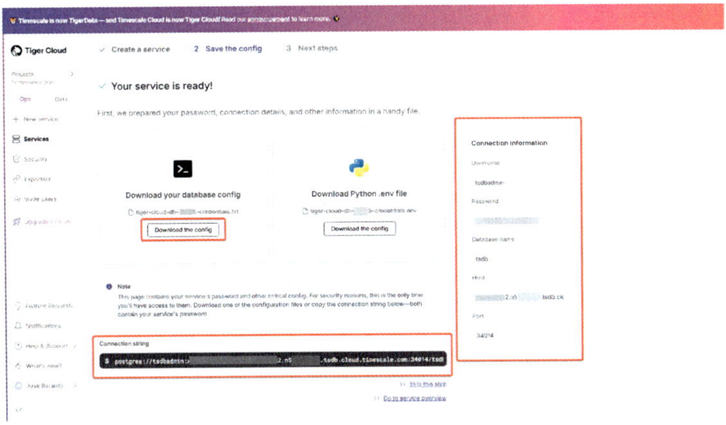

New organisation on Supabase

I'm pretty sure you know how to connect your new TigerData service from both your local psql or pgAmin 4. However, remember the config file you've downloaded?

It might surprise you with a new way of connecting our psql. If you open that file, you'll find something like:

```
echo "
[db-123456]
host=abc123456.abc123456.tsdb.cloud.timescale.com
port=34014
user=tsdbadmin
password=[random-password]
dbname=tsdb
" >> ~/.pg_service.conf
```

Surprisingly, you can run this command on both Linux, macOS or Windows Powershell. This will create a file called `pg_service.conf` under your user's home folder.

If you then just run the psql command below (replacing `123456` with your actual `DB-number`):

```
psql -d "service=db-123456"
```

You'll connect your TigerData instance using the credentials saved in the `.pg_service.conf`.

You should expect a result like this:

```
psql (17.5 (Ubuntu 17.5-1.pgdg24.04+1))
SSL connection (protocol: TLSv1.3, cipher: TLS_AES_256_GCM_SHA384, compression: off, ALPN: postgresql)
Type "help" for help.
```

```
tsdb=>
```

Connection and Testing:

```sql
-- Create a hypertable for time-series data
CREATE TABLE sensor_readings (
    time TIMESTAMPTZ NOT NULL,
    sensor_id INTEGER,
    temperature DOUBLE PRECISION,
    humidity DOUBLE PRECISION
);

-- Convert to hypertable (TigerData's core feature)
SELECT create_hypertable('sensor_readings', 'time');

-- Insert sample data
INSERT INTO sensor_readings VALUES
    (NOW(), 1, 23.5, 45.2),
    (NOW() - INTERVAL '1 hour', 1, 22.8, 44.8);

-- Use time-series optimized functions
SELECT
    time_bucket('1 hour', time) AS hour_bucket,
    sensor_id,
    AVG(temperature) as avg_temp,
    MAX(temperature) as max_temp
FROM sensor_readings
WHERE time > NOW() - INTERVAL '24 hours'
GROUP BY hour_bucket, sensor_id
ORDER BY hour_bucket DESC;
```

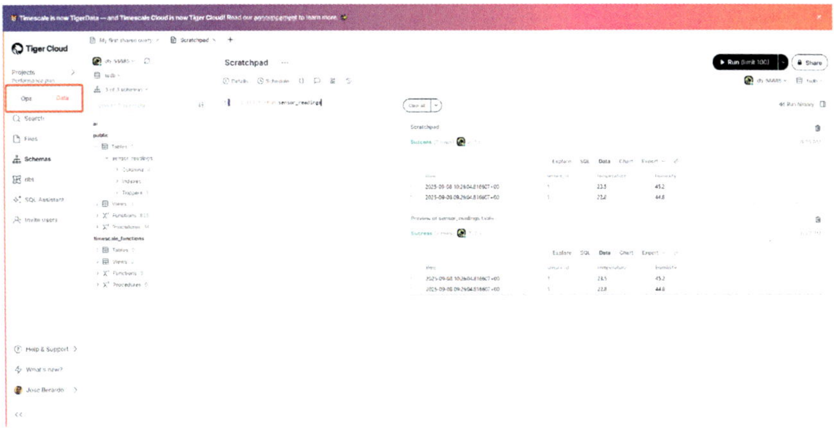

Changes applied on Tiger Cloud console

TigerData Use Cases

Modern Analytics Applications: Real-time dashboards, IoT data processing, financial market data **AI-Powered Applications**: RAG applications, vector similarity search, embedding operations **Hybrid Workloads**: Applications combining traditional transactions with time-series analytics **Developer Tools**: Monitoring systems, observability platforms, SaaS analytics

Comparing Serverless Options with Traditional Cloud

Now that you've experienced serverless PostgreSQL platforms, let's compare them with the traditional cloud services from our previous chapter:

Cost Model Comparison

Traditional Cloud (AWS RDS, Azure, GCP): - **Fixed monthly costs** for provisioned instances - **Always-on billing** whether you use the database or not
- **Manual scaling** requires intervention and planning - **Predictable but potentially wasteful** for variable workloads

Serverless Platforms: - **Usage-based pricing** - pay only for what you consume - **Scale-to-zero** when idle saves costs for development/staging - **Automatic scaling** handles traffic spikes seamlessly
- **Cost-efficient for variable workloads** and development environments

Feature and Capability Comparison

Feature	Traditional Cloud	Neon	Supabase	TigerData
Provisioning Time	5-10 minutes	Seconds	2-3 minutes	2-3 minutes

Feature	Traditional Cloud	Neon	Supabase	TigerData
Auto-scaling	Manual/Limited	Full auto-scale	Limited auto-scale	Auto-scale compute & storage
Specialized Features	Standard PostgreSQL	Database branching	Real-time APIs, Auth	Time-series, AI/Vector, Analytics
Developer Tools	Basic	SQL Editor, Monitor-ing	Full backend platform	Analytics functions, Tiger Lake
Use Case Focus	General purpose	Development workflow	Full-stack applications	Analytics, AI, time-series
Pricing Model	Fixed monthly	Usage-based	Freemium + usage	Usage-based with credits

When to Choose Each Approach

Choose Traditional Cloud When: - **Predictable, steady workloads** that run continuously - **Large enterprise applications** requiring specific compliance/control - **High-performance requirements** that need dedicated resources - **Traditional database patterns** without need for modern features

Choose Neon When: - **Development and testing workflows** that benefit from branching - **Variable workloads** that can scale to zero - **PostgreSQL-native applications** that need full compatibility - **Cost optimization** is important for development environments

Choose Supabase When: - **Full-stack applications** needing backend services - **Real-time features** are core to your application - **Rapid prototyping** and development speed matter - **Modern web applications** with authentication and APIs

Choose TigerData When: - **Time-series, analytics, or IoT data** is your primary use case - **AI and vector search capabilities** are required - **High-performance analytical queries** on large datasets - **Hybrid transactional and analytical processing** (HTAP) workloads - **Integration with data lakes** and modern analytics architectures

The Future of PostgreSQL-as-a-Service

These serverless platforms represent the direction database services are heading:

Developer-Centric Design: Features built specifically for modern development workflows **Serverless Architecture**: True pay-per-use models that scale to zero **Specialized Capabilities**: Purpose-built optimizations for specific use cases **Integrated Ecosystems**: Databases that include complementary services out of the box

Key Takeaways

Serverless is More Than Auto-Scaling: It's about developer experience, instant provisioning, and usage-based pricing **Specialized Features Matter**: Database branching, real-time APIs, and time-series optimization solve real problems **Cost Model Innovation**: Pay-per-use can be dramatically more cost-effective than always-on resources **Modern Development Patterns**: These platforms enable development workflows that weren't possible before

Looking Ahead

In our next chapter, **"Container Fundamentals,"** we'll explore another modern approach to PostgreSQL deployment using Docker and Kubernetes. While serverless abstracts infrastructure away, containers give you ultimate control and portability.

You'll see how containerized PostgreSQL complements both traditional cloud services and serverless platforms, providing yet another tool in your PostgreSQL toolkit for different scenarios and requirements.

Container Fundamentals - PostgreSQL with Docker & Kubernetes

Welcome to our exploration of containerized PostgreSQL deployment! Today we'll discover how containers revolutionize database development and deployment workflows. By the end of this chapter, you'll have hands-on experience running PostgreSQL in Docker containers and understand how Kubernetes orchestration can scale your database operations.

Containers offer something we haven't seen in our previous installation methods: **true environment consistency**, **instant provisioning**, and **effortless cleanup**. Let's see how these advantages transform your PostgreSQL experience.

Why Containers for PostgreSQL

Before diving into commands, let's understand why containers have become essential for modern PostgreSQL development and deployment.

Practical Benefits You'll Experience Today

Consistency Across Environments: The PostgreSQL container that runs on your laptop behaves identically on your colleague's machine, staging servers, and production environments. No more "it works on my machine" issues.

Instant Version Switching: Need to test your application against PostgreSQL 15, 16, 17, and 18? With containers, you can run all versions simultaneously without conflicts or complex installations.

Effortless Cleanup: Unlike traditional installations, removing a PostgreSQL container leaves no trace on your system. Perfect for testing and experimentation.

Development Workflow Enhancement: Containers integrate seamlessly with modern development practices like CI/CD pipelines, making automated testing and deployment straightforward.

Quick Demonstration: PostgreSQL in 30 Seconds

Let's see how fast containerized PostgreSQL can be. Here's what we'll accomplish in under a minute:

```
# Pull PostgreSQL image (one-time download)
docker pull postgres:17
```

```
# Start PostgreSQL with a single command
docker run --name quick-postgres -e POSTGRES_PASSWORD=demo123 -d postgres:17

# Connect and verify it's working
docker exec -it quick-postgres psql -U postgres -c "SELECT version();"
```

Compare this to the traditional installation process we covered in previous chapters - no installers, no system configuration, no service management complexity.

Introduction to Docker for PostgreSQL

Before diving into PostgreSQL on containers, let's establish a solid foundation by understanding Docker—the primary container platform we'll use throughout this chapter.

What is Docker?

Docker is an open-source platform that automates the deployment, scaling, and management of applications inside lightweight, portable containers. Think of containers as standardized shipping containers for software:

- **Containers package applications** with all their dependencies, libraries, and configuration files
- **They ensure consistent behavior** from your laptop to staging servers to production environments
- **They provide isolation** similar to virtual machines but are much more resource-efficient and faster to start

Why Docker Matters for PostgreSQL

Traditional PostgreSQL installations tie you to specific versions and configurations on each machine. With Docker:

- **Version Flexibility**: Run PostgreSQL 15, 16, 17, and 18 simultaneously without conflicts
- **Clean Environments**: Each container is isolated—no leftover files or configuration conflicts
- **Instant Provisioning**: Start a new PostgreSQL instance in seconds, not minutes
- **Consistent Environments**: The same container image works identically everywhere

Installing Docker

Desktop Installations (Recommended for Beginners)

Docker Desktop provides the easiest entry point for developers:

Docker Desktop home page

Windows 10/11: - Download Docker Desktop from docker.com - Supports WSL 2 integration for native Linux experience - Includes graphical interface for container management - Requires Windows 10 version 2004 or higher

macOS (Intel & Apple Silicon): - Native support for both Intel and M1/M2 chips - Seamless integration with macOS - Built-in Kubernetes support for advanced workflows

Installation Steps: 1. Download the installer for your platform 2. Run the installation wizard (requires admin privileges) 3. Restart your computer when prompted 4. Launch Docker Desktop and complete the initial setup

Linux Installation (Command Line)

On Linux, you typically install Docker Engine directly:

Ubuntu/Debian:

```
# Update package index
sudo apt update
sudo apt install -y apt-transport-https ca-certificates curl software-properties-common

# Add Docker's official GPG key
curl -fsSL https://download.docker.com/linux/ubuntu/gpg | sudo gpg --dearmor -o /usr/share/keyrings/docker-archive-keyring.gpg

# Add the Docker repository
echo "deb [arch=$(dpkg --print-architecture) signed-by=/usr/share/keyrings/docker-archive-keyring.gpg] https://download.docker.com/linux/ubuntu $(lsb_release -cs) stable" | sudo tee /etc/apt/sources.list.d/docker.list > /dev/null

# Update your package list with the new repository
sudo apt update
```

```
# Install Docker Engine
sudo apt install -y docker-ce docker-ce-cli containerd.io docker-buildx-plugin docker-compose-plugin

# Add your user to the docker group (optional, but recommended for running Docker without sudo)
sudo usermod -aG docker $USER
newgrp docker
```

Cloud and Hosted Options

For production workloads, cloud providers offer managed container services:

AWS: - **Amazon ECS**: Fully managed container orchestration - **AWS Fargate**: Serverless containers - **Amazon EKS**: Managed Kubernetes for complex applications

Azure: - **Azure Container Instances**: Simple container deployment - **Azure Kubernetes Service (AKS)**: Enterprise-grade orchestration

Google Cloud: - **Cloud Run**: Serverless containers with automatic scaling - **Google Kubernetes Engine (GKE)**: Managed Kubernetes

These services eliminate server management while providing scalability and high availability.

Troubleshooting Common Issues

Docker Desktop Not Starting (Windows/Mac): - Ensure virtualization is enabled in BIOS/UEFI - Check Windows features: Hyper-V and Containers are enabled - Restart Docker Desktop service

Docker Concepts You'll Need

Before jumping into PostgreSQL containers, understand these key concepts:

Images: Read-only templates containing application code and dependencies - Example: `postgres:17` contains PostgreSQL 17 with all required libraries

Containers: Running instances created from images - Multiple containers can run from the same image - Each container has its own isolated environment

Volumes: Persistent data storage that survives container deletion - Essential for database data that must persist

Networks: Communication channels between containers - Allows PostgreSQL to connect securely with application containers

Summary

For most users starting with containerized PostgreSQL:

1. **Install Docker Desktop** on Windows or Mac for integrated experience
2. **Verify installation** with version check and hello-world test
3. **Understand core concepts**: images, containers, volumes, networks

4. **Be ready for instant PostgreSQL** deployment and management

This foundation prepares you to seamlessly run PostgreSQL containers and understand the orchestration concepts we'll cover with Kubernetes later in this chapter.

Docker Prerequisites & Verification

Before we begin, let's ensure your Docker environment is ready for PostgreSQL containers.

Docker Installation Check

Step 1: Verify Docker Installation

```
# Check Docker version
docker --version

# Verify Docker is running
docker run hello-world
```

Expected Output:

```
Docker version 28.x.x, build xyz123
Hello from Docker!
```

If these commands fail, install Docker Desktop for your platform: - **Windows/Mac**: Download Docker Desktop from docker.com - **Linux**: Use your distribution's package manager or Docker's official installation script

Step 2: System Requirements Check

PostgreSQL containers need adequate resources: - **Memory**: Minimum 512MB, recommended 2GB+ for development - **Storage**: 1GB for basic PostgreSQL, more for data-heavy applications - **Network**: Port 5432 should be available (or use alternative ports)

Step 3: Docker Configuration Verification

```
# Check available disk space for Docker
docker system df

# Verify network connectivity
docker run --rm alpine ping -c 3 google.com
```

Now you're ready to run PostgreSQL containers!

Running PostgreSQL with Docker

Let's get PostgreSQL running in containers using progressively more sophisticated configurations.

Basic PostgreSQL Container

Step 1: Start Your First PostgreSQL Container

```
# Run PostgreSQL with minimal configuration
docker run --name postgres-everywhere \
  -e POSTGRES_PASSWORD=mypassword \
  -d postgres:17
```

Command Breakdown: - `--name postgres-everywhere`: Assigns a friendly name to your container - `-e POSTGRES_PASSWORD=mypassword`: Sets the superuser password (required) - `-d`: Runs container in background (detached mode) - `postgres:17`: Uses PostgreSQL version 17 official image

Step 2: Verify Container is Running

```
# Check container status
docker ps

# View container logs
docker logs postgres-everywhere
```

Successful Output:

```
CONTAINER ID   IMAGE         COMMAND                  CREATED          STATUS          PORTS      NAMES
abc123def456   postgres:17   "docker-entrypoint.s..."   30 seconds ago   Up 29 seconds   5432/tcp   postgres-everywhere

PostgreSQL init process complete; ready for start up.
LOG: database system is ready to accept connections
```

Essential Configuration with Port Mapping

Step 3: Create a Properly Configured Container

```
# Stop and remove the basic container
docker stop postgres-everywhere
docker rm postgres-everywhere

# Start with proper configuration
docker run --name pg-everywhere \
  -e POSTGRES_USER=devuser \
  -e POSTGRES_PASSWORD=set-a-strong-password \
  -e POSTGRES_DB=pg-everywhere-db \
  -p 5432:5432 \
  -d postgres:17
```

New Configuration Explained: - `-e POSTGRES_USER=devuser`: Creates a custom superuser instead of default 'postgres' - `-e POSTGRES_DB=pg-everywhere-db`: Creates a database named 'pg-everywhere-db' - `-p 5432:5432`: Maps container port 5432 to host port 5432

If you run into issues with `failed to bind host port ... tcp: address already in use`, it's because your host has a PostgreSQL already running and responding on port 5432.

In this case, you have to either shutdown the local running server, or set another port. For instance `-p 5433:5432` where `5433` is the host's port.

Step 4: Connect from Your Local Machine

```
# Connect using psql (if installed locally)
# (You don't need -p if running on default 5432)
psql -h localhost -U devuser -d pg-everywhere-db -p 5433

# Or connect through the container
docker exec -it pg-everywhere psql -U devuser -d pg-everywhere-db
```

You should get:

```
psql (17.6 (Ubuntu 17.6-1.pgdg24.04+1))
Type "help" for help.

pg-everywhere-db=#
```

Test Your Connection:

```
-- Verify PostgreSQL version and configuration
SELECT version();

-- Check current database and user
SELECT current_database(), current_user;

-- Create a test table
CREATE TABLE container_test (
    id SERIAL PRIMARY KEY,
    message TEXT,
    created_at TIMESTAMP DEFAULT CURRENT_TIMESTAMP
);

-- Insert test data
INSERT INTO container_test (message) VALUES ('Hello from PostgreSQL container!');

-- Verify data
SELECT * FROM container_test;
```

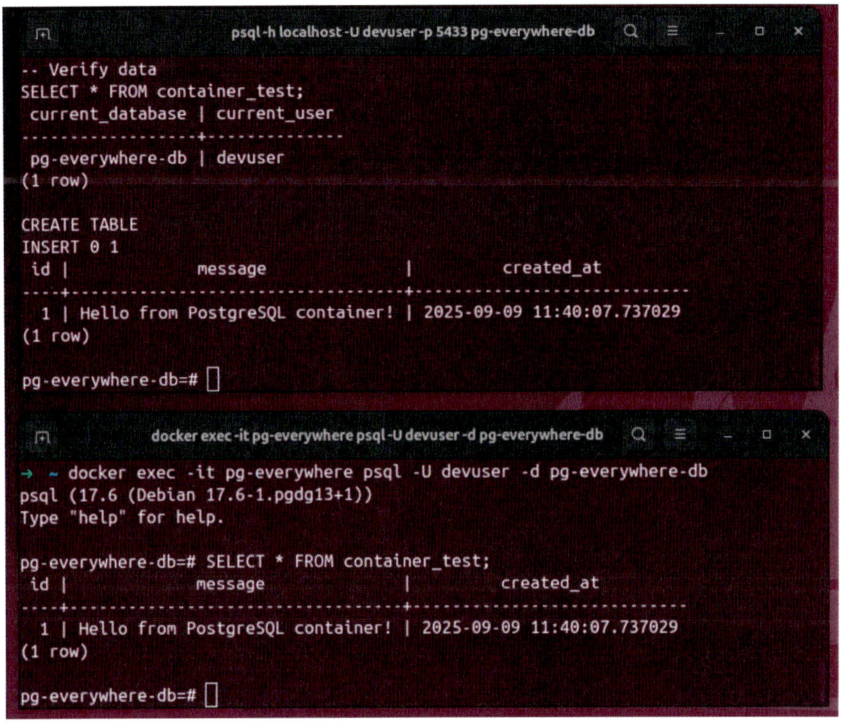

Connected to PostgreSQL through local and containerised psql

Multiple PostgreSQL Versions Side-by-Side

Step 5: Run Different PostgreSQL Versions Simultaneously

```
# PostgreSQL 14 on port 5414
docker run --name postgres-14 \
  -e POSTGRES_PASSWORD=test123 \
  -p 5414:5432 \
  -d postgres:14

# PostgreSQL 15 on port 5415
docker run --name postgres-15 \
  -e POSTGRES_PASSWORD=test123 \
  -p 5415:5432 \
  -d postgres:15

# PostgreSQL 16 on port 5416
docker run --name postgres-16 \
  -e POSTGRES_PASSWORD=test123 \
```

```
  -p 5416:5432 \
  -d postgres:16

# Check all running containers
docker ps
```

Test Different Versions:

```
# Connect to PostgreSQL 14
docker exec -it postgres-14 psql -U postgres -c "SELECT version();"

# Connect to PostgreSQL 15
docker exec -it postgres-15 psql -U postgres -c "SELECT version();"

# Connect to PostgreSQL 16
docker exec -it postgres-16 psql -U postgres -c "SELECT version();"
```

This demonstrates a key container advantage: running multiple PostgreSQL versions without conflicts!

Data Persistence & Volume Management

One critical limitation we need to address: **container data is ephemeral by default**. When you delete a container, all data disappears. Let's fix this with Docker volumes.

The Data Persistence Problem

Demonstration of the Problem:

```
# Create test data in our container
docker exec -it pg-everywhere psql -U devuser -d pg-everywhere-db -c "
CREATE TABLE important_data (id SERIAL, info TEXT);
INSERT INTO important_data (info) VALUES ('Critical business data');
SELECT * FROM important_data;"

# Stop and remove container
docker stop pg-everywhere
docker rm pg-everywhere

# Start a new container with same name
docker run --name pg-everywhere \
  -e POSTGRES_USER=devuser \
  -e POSTGRES_PASSWORD=devpass123 \
  -e POSTGRES_DB=development \
  -p 5432:5432 \
  -d postgres:17
```

```
# Check if data exists (it won't!)
docker exec -it pg-everywhere psql -U devuser -d development -c "SELECT * FROM important_data;"
```

Result: Your data is gone! This is why we need persistent volumes.

Volume Solutions

Named Volumes (Recommended Approach):

```
# Create a named volume for PostgreSQL data
docker volume create pgdata

# Run container with persistent volume
docker run --name postgres-persistent \
  -e POSTGRES_USER=devuser \
  -e POSTGRES_PASSWORD=devpass123 \
  -e POSTGRES_DB=development \
  -p 5434:5432 \
  -v pgdata:/var/lib/postgresql/data \
  -d postgres:17
```

Understanding the Volume Mount: - `-v pgdata:/var/lib/postgresql/data`: Mounts the named volume `pgdata` to PostgreSQL's data directory - `/var/lib/postgresql/data`: This is where PostgreSQL stores all database files inside the container

Data Persistence Verification

Test Data Persistence:

```
# Create important data
docker exec -it postgres-persistent psql -U devuser -d development -c "
CREATE TABLE persistent_test (
    id SERIAL PRIMARY KEY,
    test_data TEXT,
    created_at TIMESTAMP DEFAULT CURRENT_TIMESTAMP
);
INSERT INTO persistent_test (test_data) VALUES ('This data should survive container removal');
SELECT * FROM persistent_test;"

# Stop and remove the container
docker stop postgres-persistent
docker rm postgres-persistent

# Start a new container using the same volume
docker run --name postgres-recovered \
  -e POSTGRES_USER=devuser \
  -e POSTGRES_PASSWORD=devpass123 \
```

```
  -e POSTGRES_DB=development \
  -p 5434:5432 \
  -v pgdata:/var/lib/postgresql/data \
  -d postgres:17

# Verify data persisted
docker exec -it postgres-recovered psql -U devuser -d development -c "SELECT * FROM persistent_test;"
```

Success! Your data survived the container removal because it's stored in the persistent volume.

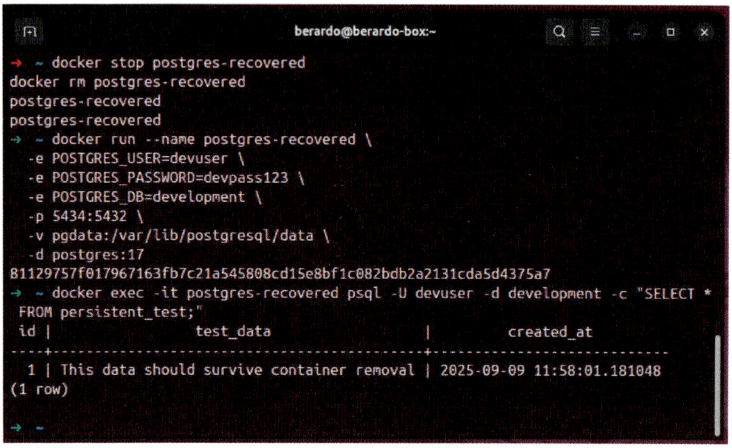

Data recovered from Docker volume

Volume Management Commands

Useful Volume Commands:

```
# List all volumes
docker volume ls

# Inspect volume details
docker volume inspect pgdata

# Create backup of volume data
docker run --rm \
  -v pgdata:/data \
  -v $(pwd):/backup \
  alpine tar czf /backup/pgdata-backup.tar.gz -C /data .

# Remove volume (when no longer needed)
```

```
docker volume rm pgdata
```

Bind Mounts (Alternative Approach)

For development, you might prefer bind mounts to store data in a specific local directory:

```
# Create local directory
mkdir -p ~/postgresql-data

# Run with bind mount
docker run --name postgres-bindmount \
  -e POSTGRES_PASSWORD=test123 \
  -p 5432:5432 \
  -v ~/postgresql-data:/var/lib/postgresql/data \
  -d postgres:17
```

Bind Mount Benefits: - Direct access to database files from host system - Easy backup using standard file system tools - Useful for development and debugging

Managing PostgreSQL Containers

Now let's master the essential container lifecycle management commands for PostgreSQL.

Container Lifecycle Management

Starting and Stopping Containers:

```
# Start a stopped container
docker start postgres-persistent

# Stop a running container gracefully
docker stop postgres-persistent

# Restart a container
docker restart postgres-persistent

# Force stop (if graceful stop fails)
docker kill postgres-persistent
```

Container Inspection and Monitoring

Monitoring Container Health:

```
# View running containers
docker ps

# View all containers (including stopped)
docker ps -a
```

```
# Monitor real-time container stats
docker stats postgres-persistent

# View detailed container information
docker inspect postgres-persistent
```

Log Management:

```
# View container logs
docker logs postgres-persistent

# Follow logs in real-time
docker logs -f postgres-persistent

# View last 50 lines of logs
docker logs --tail 50 postgres-persistent

# View logs with timestamps
docker logs -t postgres-persistent
```

Interactive Container Access

Accessing PostgreSQL Inside Containers:

```
# Execute psql commands
docker exec -it postgres-persistent psql -U devuser -d development

# Open bash shell inside container
docker exec -it postgres-persistent bash

# Run one-off commands
docker exec postgres-persistent pg_dump -U devuser development > backup.sql
```

Resource Management

Setting Resource Limits:

```
# Run PostgreSQL with resource limits
docker run --name postgres-limited \
  -e POSTGRES_PASSWORD=test123 \
  -p 5432:5432 \
  -v pgdata:/var/lib/postgresql/data \
  --memory=1g \
  --cpus=1.0 \
  -d postgres:17
```

Resource Monitoring:

```
# View resource usage
docker stats postgres-limited

# Check memory usage specifically
docker exec postgres-limited cat /proc/meminfo | head -3
```

Container Maintenance

Updating PostgreSQL Versions:

```
# Stop current container
docker stop postgres-persistent

# Pull newer PostgreSQL version
docker pull postgres:17.1

# Start with newer version (data persists via volume)
docker run --name postgres-updated \
  -e POSTGRES_USER=devuser \
  -e POSTGRES_PASSWORD=devpass123 \
  -e POSTGRES_DB=development \
  -p 5432:5432 \
  -v pgdata:/var/lib/postgresql/data \
  -d postgres:17.1
```

Cleanup Commands:

```
# Remove stopped container
docker rm postgres-old

# Remove container and associated volumes
docker rm -v postgres-temp

# Clean up unused containers, networks, and images
docker system prune
```

Kubernetes for PostgreSQL

While Docker handles individual containers beautifully, production environments often require orchestration, high availability, and automated management. That's where Kubernetes and specialized operators like CloudNativePC shine

When You Need Orchestration

Kubernetes becomes valuable for PostgreSQL when you need:

- **High Availability**: Automatic failover and recovery

- **Scaling**: Dynamic resource adjustment based on demand
- **Multi-Container Applications**: PostgreSQL integrated with application services
- **Production Operations**: Automated backups, monitoring, and maintenance

Installing Kubernetes Locally

Before deploying PostgreSQL clusters on Kubernetes, you need a Kubernetes cluster. For learning and development purposes, local Kubernetes clusters are perfect for experimentation without cloud costs.

1. Minikube (Recommended for Beginners) - Lightweight Kubernetes implementation that runs inside a VM or container - Cross-platform support (Windows, macOS, Linux) - Single command to start and stop clusters - Great learning tool with dashboard and add-ons

2. Kind (Kubernetes IN Docker) - Runs Kubernetes clusters using Docker containers as nodes - Excellent for CI/CD pipelines and automated testing - Supports multi-node clusters for advanced scenarios - Lighter weight than VM-based solutions

3. Docker Desktop Kubernetes - Built into Docker Desktop for Windows and macOS - Single-node cluster with tight Docker integration - Easiest option if you already use Docker Desktop - Good for basic Kubernetes experimentation

Installing Minikube (Step-by-Step)

Prerequisites: - **Virtualization**: Ensure virtualization is enabled in your BIOS/UEFI - **Hypervisor**: Install one of: - VirtualBox (cross-platform) - HyperKit (macOS) - Hyper-V (Windows) - **Kubectl**: Kubernetes command-line tool

Installation Steps:

1. Install kubectl:

```
# macOS (using Homebrew)
brew install kubectl

# Windows (using Chocolatey)
choco install kubernetes-cli

# Linux (using curl)
curl -LO "https://dl.k8s.io/release/$(curl -L -s https://dl.k8s.io/release/stable.txt)/bin/linux/amd64/kubectl"
sudo install -o root -g root -m 0755 kubectl /usr/local/bin/kubectl
```

2. Install Minikube:

```
# macOS
brew install minikube

# Windows
choco install minikube
```

```
# Linux
curl -Lo minikube https://storage.googleapis.com/minikube/releases/latest/minikube-linux-amd64
sudo install minikube /usr/local/bin/
```

3. Start Your Kubernetes Cluster:

```
# Start Minikube with recommended settings for PostgreSQL
minikube start --memory=4096 --cpus=2

# Verify cluster is running
kubectl cluster-info
kubectl get nodes
```

Expected Output:

```
NAME       STATUS   ROLES           AGE   VERSION
minikube   Ready    control-plane   83s   v1.34.0
```

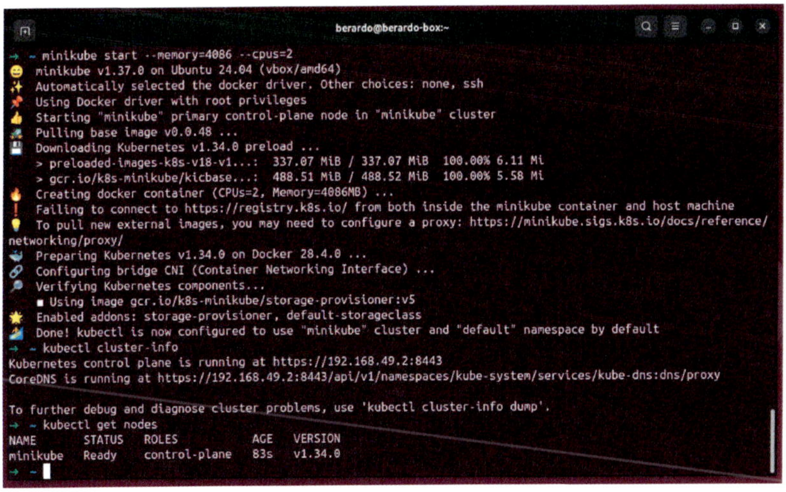

Minikube and kubectl expected command outcomes

Installing Kind (Alternative Approach)

Requirements: - Docker must be installed and running

Installation and Setup:

```
# macOS/Linux
curl -Lo ./kind https://kind.sigs.k8s.io/dl/v0.20.0/kind-linux-amd64
chmod +x ./kind
sudo mv ./kind /usr/local/bin/kind
```

```
# Windows (using Chocolatey)
choco install kind

# Create a PostgreSQL-friendly cluster
kind create cluster --name postgres-demo --config - <<EOF
kind: Cluster
apiVersion: kind.x-k8s.io/v1alpha4
nodes:
- role: control-plane
- role: worker
- role: worker
EOF

# Verify cluster
kubectl cluster-info --context kind-postgres-demo
kubectl get nodes
```

Docker Desktop Kubernetes (Simplest Option)

Enable Kubernetes: 1. Open Docker Desktop 2. Go to Settings → Kubernetes 3. Check "Enable Kubernetes" 4. Click "Apply & Restart" 5. Wait for Kubernetes to start (green indicator)

Verify Installation:

```
kubectl config current-context
# Should show: docker-desktop

kubectl get nodes
# Should show single docker-desktop node
```

Basic Kubernetes PostgreSQL Deployment

Prerequisites Check:

```
# Verify kubectl is installed and working
kubectl version --client

# Check cluster connection
kubectl get nodes

# Verify you have administrative access
kubectl auth can-i create namespace
```

If you don't have a Kubernetes cluster yet, choose one of the installation methods above.

CloudNativePG Introduction

CloudNativePG is a **PostgreSQL-specific Kubernetes operator** that brings database expertise to Kubernetes orchestration. It's not just about running containers—it's about running PostgreSQL **properly** in Kubernetes.

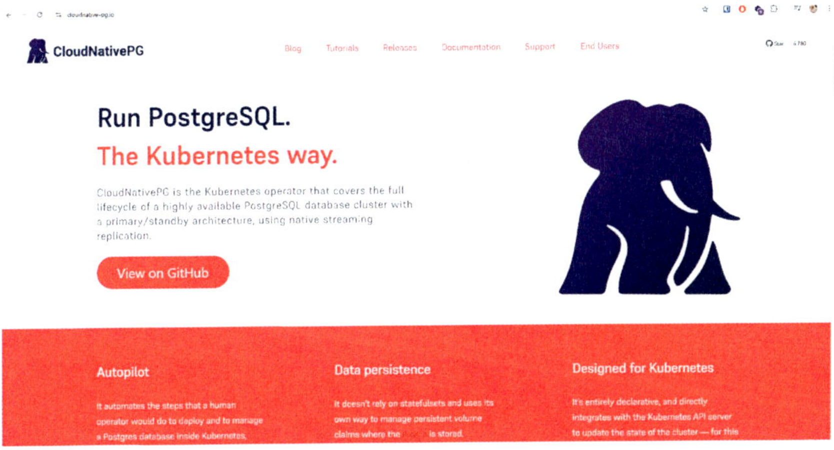

CloudNativePG project page

Key CloudNativePG Features: - **Automated High Availability**: Primary/standby architecture with automatic failover - **Backup Management**: Automated backups to cloud storage or volume snapshots - **Rolling Updates**: Seamless PostgreSQL version upgrades - **Built-in Monitoring**: Prometheus metrics and observability - **Declarative Configuration**: Pure Kubernetes YAML—no additional tools required

Basic Kubernetes PostgreSQL Deployment

Prerequisites Check:

```
# Verify kubectl is installed
kubectl version --client

# Check cluster connection (if using minikube, kind, or cloud cluster)
kubectl get nodes
```

If you don't have a Kubernetes cluster, you can set up a local one:

```
# Using minikube (easiest for testing)
minikube start

# Or using kind (Docker-based)
kind create cluster --name postgres-demo
```

Installing CloudNativePG

Step 1: Install the CloudNativePG Operator:

```
# Install CloudNativePG operator
kubectl apply -f https://raw.githubusercontent.com/cloudnative-pg/cloudnative-pg/release-1.20/releases/cnpg-1.20.0.yaml

# Verify operator is running
kubectl get pods -n cnpg-system
```

Expected Output:

```
NAME                              READY   STATUS    RESTARTS   AGE
cnpg-controller-manager-xxx-xxx   1/1     Running   0          17s
```

Deploying a PostgreSQL Cluster

Step 2: Create a PostgreSQL Cluster:

Create a file named `postgres-cluster.yaml`:

```
apiVersion: postgresql.cnpg.io/v1
kind: Cluster
metadata:
  name: postgres-cluster
spec:
  instances: 3

  postgresql:
    parameters:
      max_connections: "200"
      shared_buffers: "256MB"
      effective_cache_size: "1GB"

  bootstrap:
    initdb:
      database: webapp
      owner: appuser
      secret:
        name: postgres-credentials

  storage:
    size: 1Gi
    storageClass: standard

---
apiVersion: v1
kind: Secret
```

```
metadata:
  name: postgres-credentials
type: kubernetes.io/basic-auth
stringData:
  username: appuser
  password: myapp123
```

Step 3: Deploy the Cluster:

```
# Apply the configuration
kubectl apply -f postgres-cluster.yaml

# Watch cluster creation (ctrl + c to stop watching)
kubectl get cluster postgres-cluster -w
```

Step 4: Verify Deployment:

```
# Check cluster status
kubectl get cluster postgres-cluster

# View all pods in the cluster
kubectl get pods -l postgresql=postgres-cluster
```

Expected Output:

```
NAME              AGE    INSTANCES   READY   STATUS                     PRIMARY
postgres-cluster  5m     3           3       Cluster in healthy state   postgres-cluster-1


NAME                 READY   STATUS    RESTARTS   AGE
postgres-cluster-1   1/1     Running   0          5m
postgres-cluster-2   1/1     Running   0          5m
postgres-cluster-3   1/1     Running   0          4m
```

Connecting to Your Kubernetes PostgreSQL

Step 5: Connect to the Database:

```
# Check services created
kubectl get services

# Get the service connection details
kubectl get services postgres-cluster-rw

# Port-forward to access locally (for testing - avoid port conflicts)
kubectl port-forward service/postgres-cluster-rw 5440:5432

# In another terminal, connect using psql
```

```
psql -h localhost -U appuser -d webapp -p 5440
```

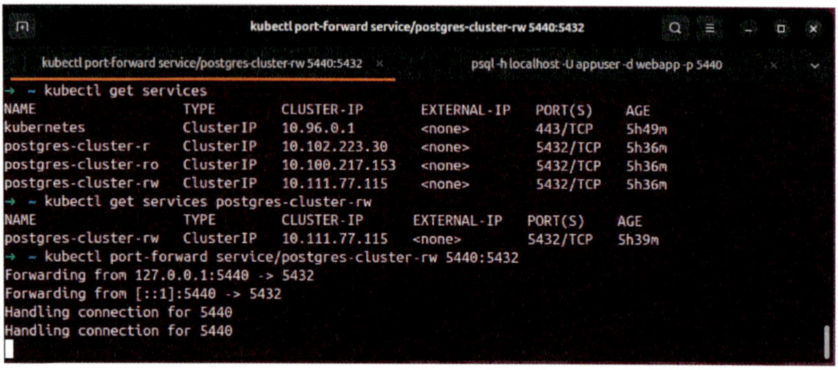

postgres-cluster terminal

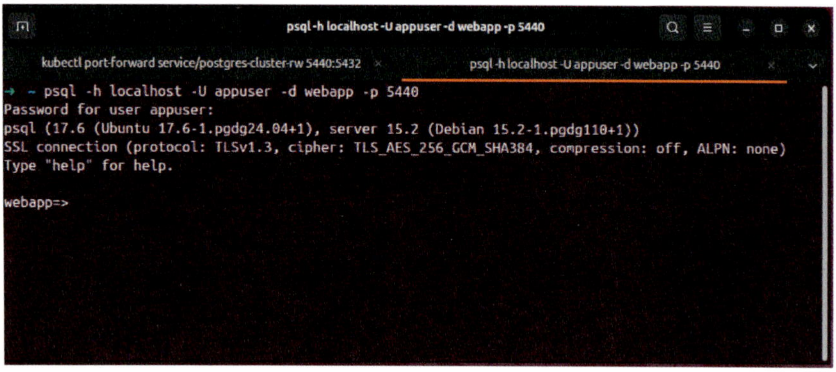

psql connected to postgres-cluster-rw terminal

Step 6: Test High Availability:

```
# Identify the primary instance
kubectl get cluster postgres-cluster -o yaml | grep targetPrimary

# Delete the primary pod to trigger failover
kubectl delete pod postgres-cluster-1

# Watch automatic failover
kubectl get pods -l postgresql=postgres-cluster -w
```

CloudNativePG automatically promotes a standby to primary and maintains service availability!

Essential kubectl Commands for PostgreSQL

Cluster Management:

```
# View cluster details
kubectl describe cluster postgres-cluster

# Check cluster logs
kubectl logs -l postgresql=postgres-cluster

# Scale cluster (add read replicas)
kubectl patch cluster postgres-cluster --type='merge' -p='{"spec":{"instances":5}}'

# Backup cluster
kubectl apply -f - <<EOF
apiVersion: postgresql.cnpg.io/v1
kind: Backup
metadata:
  name: backup-$(date +%Y%m%d%H%M%S)
spec:
  cluster:
    name: postgres-cluster
EOF
```

Monitoring and Troubleshooting:

```
# Monitor cluster status
kubectl get cluster postgres-cluster -o wide

# View cluster events
kubectl get events --field-selector involvedObject.name=postgres-cluster

# Access PostgreSQL configuration
kubectl exec postgres-cluster-1 -- cat /var/lib/postgresql/data/postgresql.conf
```

Container Best Practices for PostgreSQL

As we conclude, let's establish essential practices for running PostgreSQL in containers professionally.

Security Essentials

Use Non-Root Users:

```
# Run PostgreSQL container as non-root user
docker run --name postgres-secure \
  --user 999:999 \
  -e POSTGRES_PASSWORD=securepass \
```

```
  -v pgdata:/var/lib/postgresql/data \
  -d postgres:17
```

Manage Secrets Properly:

```
# Use Docker secrets instead of environment variables
echo "mysecretpassword" | docker secret create pg_password -

# Use secrets in Docker Swarm
docker service create \
  --name postgres \
  --secret pg_password \
  -e POSTGRES_PASSWORD_FILE=/run/secrets/pg_password \
  postgres:17
```

Network Security:

```
# Create custom network for PostgreSQL
docker network create postgres-network

# Run PostgreSQL on custom network
docker run --name postgres-networked \
  --network postgres-network \
  -e POSTGRES_PASSWORD=networkpass \
  -d postgres:17

# Connect application to same network
docker run --name app \
  --network postgres-network \
  your-app:latest
```

Production Readiness

Health Checks:

```
# Add health check to PostgreSQL container
docker run --name postgres-health \
  -e POSTGRES_PASSWORD=healthpass \
  --health-cmd="pg_isready -U postgres" \
  --health-interval=30s \
  --health-timeout=10s \
  --health-retries=3 \
  -d postgres:17
```

Resource Limits and Monitoring:

```
# Production-ready container with limits
docker run --name postgres-production \
```

```
  -e POSTGRES_PASSWORD=prodpass \
  -v pgdata:/var/lib/postgresql/data \
  --memory=2g \
  --cpus=2.0 \
  --restart=unless-stopped \
  --health-cmd="pg_isready -U postgres" \
  -p 5432:5432 \
  -d postgres:17
```

Logging Configuration:

```
# Configure structured logging
docker run --name postgres-logged \
  -e POSTGRES_PASSWORD=logpass \
  -e POSTGRES_INITDB_ARGS="--auth-host=md5" \
  --log-driver=json-file \
  --log-opt max-size=100m \
  --log-opt max-file=3 \
  -d postgres:17
```

Version Management

Image Tagging Strategy:

```
# Use specific versions, not 'latest'
docker pull postgres:17.1
docker pull postgres:16.4
docker pull postgres:15.8

# Tag for your environment
docker tag postgres:17.1 mycompany/postgres:17.1-prod
docker tag postgres:17.1 mycompany/postgres:latest
```

CI/CD Pipeline Integration

Automated Testing Example:

```
# Start PostgreSQL for testing
docker run --name test-postgres \
  -e POSTGRES_PASSWORD=testpass \
  -e POSTGRES_DB=testdb \
  -p 5432:5432 \
  -d postgres:17

# Wait for PostgreSQL to be ready
until docker exec test-postgres pg_isready -U postgres; do
  echo "Waiting for PostgreSQL..."
```

```
  sleep 2
done

# Run your tests
# pytest tests/ --postgresql-url=postgresql://postgres:testpass@localhost:5432/testdb

# Cleanup after tests
docker stop test-postgres
docker rm test-postgres
```

This pattern integrates perfectly with CI/CD systems like GitHub Actions, GitLab CI, or Jenkins.

PostgreSQL on Podman: A Viable Docker Alternative

Podman is designed as a drop-in replacement for Docker with OCI (Open Container Initiative) compatibility and PostgreSQL containers work identically on both platforms.

Running PostgreSQL with Podman

Basic PostgreSQL Container with Podman:

```
# Pull PostgreSQL image (from Docker Hub or other registries)
podman pull docker.io/library/postgres:17

# Run PostgreSQL container
podman run --name postgres-podman \
  -e POSTGRES_PASSWORD=mypassword \
  -e POSTGRES_DB=testdb \
  -p 5432:5432 \
  -v pgdata:/var/lib/postgresql/data \
  -d postgres:17

# Connect to the database
podman exec -it postgres-podman psql -U postgres -d testdb
```

Commands are nearly identical to Docker - simply replace `docker` with `podman` in most cases.

Benefits of Podman over Docker for PostgreSQL

1. Enhanced Security Architecture - **Daemonless operation**: No background daemon running with root privileges - **Rootless containers by default**: PostgreSQL runs without elevated system privileges - **Better audit trails**: Container processes retain user session information for security auditing

2. Resource Efficiency - **No daemon overhead**: Eliminates Docker daemon resource consumption - **Faster container startup**: Direct process execution without daemon intermediation - **Lower memory footprint**: No persistent background services

3. System Integration - **Native systemd integration**: Generate systemd service files directly from containers - **Better process management**: Container processes appear as regular system processes - **Simplified monitoring**: Standard Linux tools can monitor container processes directly

Drawbacks Compared to Docker

1. Ecosystem Maturity - **Smaller community**: Less extensive documentation and community support - **Fewer integrations**: Some third-party tools expect Docker-specific features - **Platform limitations**: Windows support is more complex than Docker Desktop

2. Development Experience - **Limited GUI options**: No equivalent to Docker Desktop's integrated experience - **Compose limitations**: `podman-compose` is less mature than `docker-compose` - **Tool compatibility**: Some Docker-centric development tools may require adaptation

3. Cloud Service Integration - **Service expectations**: Some cloud platforms optimize for Docker runtime specifics - **Registry compatibility**: While broadly compatible, some registries may have Docker-specific features

Kubernetes Integration with Podman

Podman and Kubernetes Compatibility: - Podman containers work seamlessly in Kubernetes clusters - Kubernetes typically uses containerd or CRI-O as runtime, both compatible with OCI standards - Podman's local "pod" concept aligns well with Kubernetes pod architecture

Local Development Workflow:

```
# Create a pod with multiple containers (like Kubernetes)
podman pod create --name postgres-pod -p 5432:5432

# Run PostgreSQL in the pod
podman run -d --pod postgres-pod \
  -e POSTGRES_PASSWORD=password \
  --name postgres-db postgres:17

# Run pgAdmin in the same pod
podman run -d --pod postgres-pod \
  -e PGADMIN_DEFAULT_EMAIL=admin@example.com \
  -e PGADMIN_DEFAULT_PASSWORD=admin \
  --name pgadmin dpage/pgadmin4
```

CloudNativePG with Podman-Based Workflows

CloudNativePG Compatibility: - CloudNativePG operates at the Kubernetes orchestration level - **Runtime agnostic**: Works regardless of whether nodes use Docker, containerd, or Podman-compatible runtimes - **Standard PostgreSQL images**: Uses the same official PostgreSQL container images that work with Podman

Production Deployment Workflow: 1. **Local Development**: Use Podman for secure, rootless

PostgreSQL development 2. **Testing**: Deploy the same container images to Kubernetes with CloudNativePG 3. **Production**: CloudNativePG manages high availability, backups, and scaling automatically

Example CloudNativePG Integration:

```yaml
apiVersion: postgresql.cnpg.io/v1
kind: Cluster
metadata:
  name: postgres-cluster
spec:
  instances: 3

  # Same PostgreSQL image that works with Podman
  imageName: postgres:17

  postgresql:
    parameters:
      max_connections: "200"
      shared_buffers: "256MB"

  bootstrap:
    initdb:
      database: webapp
      owner: appuser

  storage:
    size: 10Gi
```

When to Choose Podman for PostgreSQL

Ideal Use Cases: - **Security-conscious environments**: Where rootless operation is mandatory - **Resource-constrained systems**: Where daemon overhead matters - **SystemD-centric infrastructure**: Linux environments prioritizing native service management - **Development workflows**: Where quick container startup and cleanup matter

Docker Might Be Better When: - **Team standardization**: Existing Docker expertise and toolchain - **Complex orchestration**: Heavy reliance on Docker Compose or Swarm - **Windows development**: Where Docker Desktop integration is crucial - **Third-party integrations**: Tools specifically designed for Docker

Practical Migration Path

Moving from Docker to Podman for PostgreSQL: 1. **Test compatibility**: Run existing PostgreSQL containers with Podman 2. **Update scripts**: Replace `docker` commands with `podman` equivalents 3. **Adapt compose files**: Convert to `podman-compose` or Kubernetes manifests 4. **Production deployment**: Use CloudNativePG for Kubernetes orchestration regardless of local development runtime

Summary: Podman offers a more secure, resource-efficient alternative to Docker for PostgreSQL workloads, especially in security-conscious or resource-constrained environments. While Docker maintains broader ecosystem support, Podman's advantages in security and system integration make it increasingly attractive for PostgreSQL deployments. CloudNativePG provides the same robust PostgreSQL orchestration capabilities regardless of your container runtime choice, making the transition path flexible and non-disruptive.

Key Takeaways and Next Steps

Congratulations! You've mastered PostgreSQL containers from basic Docker usage to production Kubernetes deployments. Let's summarize what you've accomplished:

Container Mastery: - Started PostgreSQL containers with proper configuration using Docker (later discussed Podman as an alternative) - Implemented data persistence with volumes - Managed multiple PostgreSQL versions simultaneously - Applied security and resource management best practices

Kubernetes Understanding: - Deployed high-availability PostgreSQL clusters with CloudNativePG - Understood when orchestration adds value - Experienced automatic failover and recovery - Learned essential kubectl commands for database operations

Production Readiness: - Applied security best practices - Configured monitoring and health checks - Established version management strategies - Designed CI/CD integration patterns

Connecting to Your PostgreSQL Journey

These containerization skills directly enhance every other topic in this course:

SQL Development: Use containers to test queries across different PostgreSQL versions **Performance Tuning**: Easily spin up test environments for optimization experiments **High Availability**: CloudNativePG provides the foundation for enterprise-grade deployments **Backup and Recovery**: Container-based strategies integrate with modern infrastructure

What's Coming Next

In our final chapter, **"PostgreSQL Tools & Interfaces,"** we'll explore psql (more?!), pgAdmin, and other essential tools for database management. You'll see how these tools integrate beautifully with containerized PostgreSQL instances, providing powerful ways to interact with your databases regardless of how they're deployed.

PostgreSQL Tools & Interfaces

Now, let's have an overview of the essential tools and interfaces that make working with PostgreSQL both powerful and efficient. In our journey so far, we've covered installation across different platforms, and now it's time to master the tools you'll use daily.

Understanding PostgreSQL Connections

First and foremost, any tool will need to connect to a database instance. Therefore, let me introduce you the two primary connection methods supported by PostgreSQL:

- Unix-domain sockets (local connections) and
- TCP/IP sockets (network connections).

Unix-Domain Sockets Unix-domain sockets provide fast and secure local connections on Unix-like systems. These connections are: - Only available on the same machine - More efficient than TCP/IP for local connections - Default connection method on Unix-like systems

TCP/IP Connections TCP/IP connections allow remote access to PostgreSQL across networks. These connections: - Work across networks and the internet - Require proper firewall and security configuration - Are necessary for cloud database access

Understanding these connection types is crucial when working with PostgreSQL, especially as you move towards production environments and cloud deployments.

PostgreSQL Connection Configuration

Two main configuration files control PostgreSQL connection settings: - postgresql.conf - pg_hba.conf

postgresql.conf

postgresql.conf plays a crucial role in controlling database connections and server behavior. Here are the key connection-related settings:

- `listen_addresses`: Controls which IP addresses PostgreSQL listens on
- `port`: Specifies the TCP port (default 5432)
- `max_connections`: Maximum number of concurrent connections allowed
- `unix_socket_directories`: Location for Unix-domain socket files

Example postgresql.conf connection settings:

```
listen_addresses = '*'      # Listen on all available IP addresses
port = 5432                 # Default PostgreSQL port
max_connections = 100       # Maximum concurrent connections
```

While postgresql.conf controls how the server accepts connections, it works in tandem with pg_hba.conf which controls who can connect, so let's go ahead and check this file now.

pg_hba.conf

The pg_hba.conf file controls client authentication and is one of PostgreSQL's primary security mechanisms. Think of it as your database's bouncer - it decides who gets in and how they need to identify themselves.

Purpose and Location - Located in the database cluster's data directory - Controls which users can connect from which hosts - Defines authentication methods for different connection types

File Structure Each line in pg_hba.conf contains up to five fields:

```
# TYPE  DATABASE        USER            ADDRESS                 METHOD

# "local" is for Unix domain socket connections only
local   all             all                                     peer
# IPv4 local connections:
host    all             all             127.0.0.1/32            trust
# IPv6 local connections:
host    all             all             ::1/128                 scram-sha-256
```

Field 1: TYPE (Connection Type) - `local`: Unix-domain socket connections (ADDRESS field is always empty for local connections) - `host`: TCP/IP connections (both IPv4 and IPv6) - `hostssl`: SSL-encrypted TCP/IP connections only - `hostnossl`: Non-SSL TCP/IP connections only - `hostgssenc`: GSSAPI-encrypted TCP/IP connections only - `hostnogssenc`: Non-GSSAPI-encrypted TCP/IP connections only

Field 2: DATABASE (Target Database) - `all`: All databases - Specific database name(s): `pg-everywhere`, `production` - `sameuser`: Database with the same name as the connecting user - `samerole`: Databases where the user is a member of the role - `replication`: For physical replication connections - Multiple databases can be specified separated by commas

Field 3: USER (Database User) - `all`: All users - Specific username(s): `postgres`, `appuser` - Group roles with `+` prefix: `+admin_group` (indicates role membership) - Multiple users can be specified separated by commas

Field 4: ADDRESS (Client Address) - Left blank for `local` connections - IP address or CIDR range: `192.168.1.0/24`, `10.0.0.100/32` - Special keywords: `all`, `samehost`, `samenet` - Hostname: `myserver.company.com` - `0.0.0.0/0` represents all IPv4 addresses

Field 5: METHOD (Authentication Method) - `trust`: Allow connection unconditionally (no password required) - (not secure - avoid in production) - `reject`: Reject connection unconditionally - `md5`: MD5-encrypted password authentication (legacy) - `scram-sha-256`: Modern secure password authentication (recommended) - `password`: Plaintext password (not secure - avoid) - `peer`: Use operating system user identity for authentication - `ident`: Contact ident server on client for user verification - External methods:

`gss`, `sspi`, `ldap`, `radius`, `cert`, `pam`

Common Use Cases - Allowing local connections without password: `local all all trust` - Restricting remote access to specific IP ranges: `host mydb myuser 192.168.1.0/24 md5` - Enforcing SSL for remote connections: `hostssl all all 0.0.0.0/0 scram-sha-256` - Denying specific problematic hosts: `host all all 10.0.1.100/32 reject`

Important Notes - PostgreSQL processes these rules sequentially from top to bottom - The **first matching rule** is used - there's no fall-through to subsequent rules - Order matters: place more specific rules before general ones - Changes require at least a configuration reload (`pg_reload_conf()` or service restart)

This field-by-field structure gives you precise control over who can connect to your PostgreSQL server, from where, and how they authenticate.

Configuration Management Essentials

pg_ctl command and pg_ctlcluster

The traditional PostgreSQL command line control utility is `pg_ctl`, used to start, stop, restart, reload, and check status of the server process.

- On Debian-based Linux systems (including Ubuntu), `pg_ctl` has been largely replaced by a wrapper script called `pg_ctlcluster`. This wrapper manages multiple PostgreSQL instances (clusters) and handles service requests from the system's init scripts.
- When you use commands like `systemctl start postgresql` or `systemctl reload postgresql` on Linux, the service management system invokes `pg_ctlcluster` internally to handle those operations.
- Similarly, on macOS, `brew services` commands manage PostgreSQL services by interfacing with the standard PostgreSQL `pg_ctl` behind the scenes for controlling start, stop, and reload operations.

pg_reload_conf() function - The Windows Solution

- PostgreSQL provides the SQL function `pg_reload_conf()`, which allows you to reload configuration files from within a running database session.
- This function sends a SIGHUP signal to the server process, instructing it to re-read its configuration files (`postgresql.conf` and `pg_hba.conf`).
- **Importantly**, this works on all platforms, including Windows, where traditional reload commands like `pg_ctl reload` aren't available through service management tools.
- Using `SELECT pg_reload_conf();` is a platform-independent way to apply configuration changes that support reload without disrupting active connections.

Configuration reload vs restart behavior

PostgreSQL parameters are categorized by their context, which determines whether changes can be applied via reload or require a full server restart:

Parameters requiring restart (context = 'postmaster'): - Must be set at server startup and cannot be changed without restarting - Include fundamental server settings like network configuration and core memory allocation - Examples: `listen_addresses`, `port`, `max_connections`, `shared_buffers`, `wal_level`

Parameters supporting reload (context = 'sighup', 'user', 'superuser'): - Can be re-read and applied without restarting the server - Include logging settings, autovacuum configuration, and many runtime parameters - Examples: `log_destination`, `log_directory`, `log_min_duration_statement`, `autovacuum`

Special case - pg_hba.conf: - Changes to `pg_hba.conf` (authentication rules) **always** take effect after reload - No restart required regardless of the specific authentication changes made - New connections immediately use the updated authentication rules

Summary of applying configuration changes

Action	Applies To	Effect on Connections	Requires Restart?
Reload	`postgresql.conf`, `pg_hba.conf`	Existing connections stay active; new connections use updated config	Only for certain parameters
Restart	All parameters	All connections terminated and must reconnect	Required for `postmaster` context params

Parameters Requiring Full Server Restart (`context = 'postmaster'`)

Parameter	Purpose	Example Value
`listen_addresses`	Network interfaces to listen on	`'*'` or `'localhost'`
`port`	TCP port number	`5432`
`max_connections`	Maximum concurrent connections	`200`
`shared_buffers`	Memory for caching database pages	`512MB`
`wal_level`	Write-ahead logging detail level	`logical`
`max_worker_processes`	Maximum background worker processes	`16`
`max_prepared_transactions`	Maximum prepared transactions	`100`

Parameters Supporting Reload (No Restart Required)

Parameter	Purpose	Example Value
`log_destination`	Where to send log output	`'stderr'`
`log_directory`	Directory for log files	`'pg_log'`

Parameter	Purpose	Example Value
`log_min_duration_statement`	Log slow queries (milliseconds)	`1000`
`autovacuum`	Enable automatic vacuuming	`on`
`autovacuum_naptime`	Time between autovacuum runs	`1min`
`work_mem`	Memory for query operations	`4MB`
`maintenance_work_mem`	Memory for maintenance operations	`64MB`

How to Check Parameter Requirements

You can query which parameters require restart:

```
-- Parameters requiring restart
SELECT name, context, short_desc
FROM pg_settings
WHERE context = 'postmaster'
ORDER BY name;

-- Parameters supporting reload
SELECT name, context, short_desc
FROM pg_settings
WHERE context IN ('sighup', 'user', 'superuser')
ORDER BY name;
```

Practical Implications

For Linux/macOS: - Use `reload` for most logging, autovacuum, and memory settings - Use `restart` for connection, port, and core memory allocation changes

For Windows: - Generally plan for service restarts when making configuration changes - Consider using the SQL `SELECT pg_reload_conf();` approach for reload-supported parameters if you have an active connection

Command-Line Interface: psql

Let's start our PostgreSQL tools journey with psql, PostgreSQL's native command-line interface. While it might seem intimidating at first, psql is incredibly powerful and often the fastest way to interact with your database.

Connection Methods

Local Socket Connections For local socket connections:

```
# Connect to user's default database
psql -U postgres
# Connect to a specific database using URI format
psql "postgresql:///mydatabase"
```

TCP/IP Connections For TCP/IP connections:

```
# Connect using individual parameters
psql -h localhost -p 5432 -U postgres
# Connect using URI format
psql "postgresql://postgres@localhost:5432/mydatabase"
```

Understanding the psql Prompt

When you launch psql, you're greeted with a prompt that provides valuable information about your session state. The basic format is:

```
database_name[session_state_indicator][transaction_state_indicator][privilege_level]
```

Session State Indicators The first special character reflects your current command state:

- `=` Normal state (ready for new command)
- `-` Multi-line command continuation
- `'` Inside single-quoted string that spans multiple lines
- `"` Inside double-quoted string that spans multiple lines
- `)` Unmatched parentheses or brackets

Transaction State Indicators An additional character that appears when inside transactions:

- `*` Inside a transaction block (after BEGIN)
- `!` Inside a failed transaction block (must be rolled back)

Privilege Level Indicators The final character shows your user privileges: - # Superuser privileges - > Regular user privileges

Examples of Different Prompts

Normal Operations:

```
postgres=#      # Superuser, ready for command
myapp=>         # Regular user, ready for command
sales=#         # Superuser, ready for command
```

Multi-line Commands:

```
postgres-# SELECT id, name, email
postgres-# FROM users
postgres-# WHERE active = true;
```

String Continuation:

```
postgres=# INSERT INTO products (description) VALUES ('This is a very long
postgres'# product description that spans
postgres'# multiple lines in our database');
```

Unmatched Parentheses:

```
postgres=# SELECT COUNT(
postgres(# id
postgres(# ) FROM users;
```

Transaction States:

```
postgres=# BEGIN;
postgres=*# SELECT * FROM users;         -- Inside transaction block
postgres=*# INSERT INTO users VALUES ('John', 'john@email.com');
postgres=*# COMMIT;
postgres=# -- Back to normal state

postgres=# BEGIN;
postgres=*# SELECT * FROM nonexistent_table;
ERROR:  relation "nonexistent_table" does not exist
postgres=!# SELECT * FROM users;       # Failed transaction, must rollback
ERROR:  current transaction is aborted, commands ignored until end of transaction block
postgres=!# ROLLBACK;
postgres=# -- Back to normal state
```

Complex Combined States:

```
postgres=# BEGIN
postgres-# ;  -- Multi-line BEGIN command
BEGIN
postgres=*# CREATE TABLE test(  -- Inside transaction, unmatched parenthesis
postgres(*# id INTEGER,
postgres(*# name TEXT
postgres(*# );
CREATE TABLE
postgres=*# INSERT INTO test VALUES ('Long string
postgres'*# that spans multiple
postgres'*# lines' -- Parenthesis continuation inside transaction
postgres'*# ); -- String continuation inside transaction
INSERT 0 1
postgres=*# COMMIT;
COMMIT
postgres=# -- Back to normal
```

Failed Transaction with Complex States:

```
postgres=# BEGIN;
```

```
postgres=*# SELECT * FROM wrong_table;
ERROR:  relation "wrong_table" does not exist
postgres=!# CREATE TABLE users( -- Failed transaction with unmatched parenthesis
postgres(!# id INTEGER,
postgres(!# name TEXT
postgres(!# );
ERROR:  current transaction is aborted, commands ignored until end of transaction block
postgres=!# INSERT INTO users VALUES ('Name with
postgres'!# multiple lines');  -- String continuation in failed transaction
ERROR:  current transaction is aborted, commands ignored until end of transaction block
postgres=!# ROLLBACK;
ROLLBACK
postgres=#  -- Back to normal
```

Essential Slash Commands

Getting Help

```
\?                    -- Show all psql commands
\h                    -- Show help for SQL commands
\h CREATE             -- Show specific help for all CREATE commands
\h CREATE DATABASE    -- Show specific help for CREATE DATABASE command
```

Navigation and Connection

```
\l                    -- List all databases
\c                    -- Show current connection info
\c mydatabase         -- Connect to specified database
\c mydatabase user    -- Connect as different user
\q                    -- Quit psql
```

Object Information Commands

```
\d                    -- List tables, views, and sequences
\d+                   -- Same as \d with additional details
\d tablename          -- Show table structure
\d+ tablename         -- Detailed table structure with descriptions
\dt                   -- List tables only
\dv                   -- List views only
\df                   -- List functions
\di                   -- List indexes
\du                   -- List users and roles
```

Query Management Commands

Query Buffer Operations

```
\e                  -- Edit command buffer in external editor
\g                  -- Execute previous command
\s                  -- Show command history
\w filename         -- Write command history to file
\r                  -- Reset query buffer (clear current command)
```

Display and Output Control

```
\timing             -- Toggle query execution time display
\x                  -- Toggle expanded display (useful for wide tables)
\pset border 2      -- Set table border style
\o filename         -- Send query results to file
\o                  -- Reset output to terminal
```

File Operations

Executing Scripts

```
\i filename         -- Execute commands from a file
\ir filename        -- Execute commands relative to current script
```

Working with Variables

```
\set                -- Show all variables
\set varname value  -- Set a variable
\unset varname      -- Unset a variable
```

Practical Usage Tips

Escaping Multi-line Mode If you find yourself stuck in any multi-line mode, you have several options:

1. **Complete the statement properly** - Add the missing quote, parenthesis, or semicolon
2. **Use \r to reset the query buffer:**

   ```
   postgres'*# \r
   postgres=*#
   ```

3. **Use Ctrl+C** - Cancels the current command and returns to normal prompt

Using Variables in Queries

```
\set user_id 123
SELECT * FROM users WHERE id = :user_id;
```

Understanding Prompt Combinations The key insight is that transaction indicators (`*` and `!`) are **additional** characters that combine with session state indicators:

- `postgres=*#` - Normal state inside transaction
- `postgres-*#` - Multi-line continuation inside transaction
- `postgres'*#` - String continuation inside transaction

- `postgres(*#` - Unmatched parentheses inside transaction
- `postgres=!#` - Normal state in failed transaction
- `postgres'!#` - String continuation in failed transaction

These commands form the foundation of efficient database management through psql. While GUI tools are helpful, mastering these commands will significantly improve your productivity when working with PostgreSQL. Think of these commands as your Swiss Army knife - they're essential tools that become second nature with practice.

pgAdmin 4: Visual Database Management

Now let's transition to pgAdmin, PostgreSQL's official graphical interface. While psql excels at rapid commands and scripting, pgAdmin provides a comprehensive visual environment for database management.

As you'll recall from the visual installation process, pgAdmin 4 was included as part of the PostgreSQL installation package. Now it's time to explore its powerful features that make database administration more intuitive and efficient.

If you decided not to locally install PostgreSQL, or just used Postgres.app on macOS, you can download and install pgAdmin 4 separately at pgadmin.org. We're not going to cover this installation because, at this stage, it's super trivial.

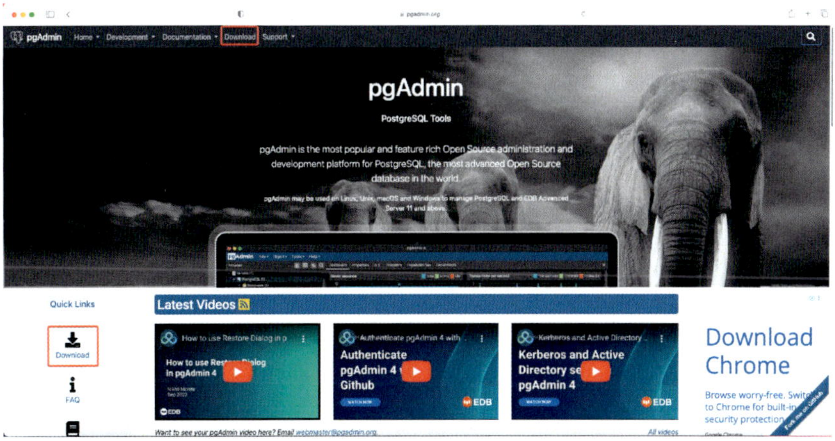

pgAdmin homepage

Interface Overview

When you first launch pgAdmin 4, you'll notice its clean, organized interface designed for productivity:

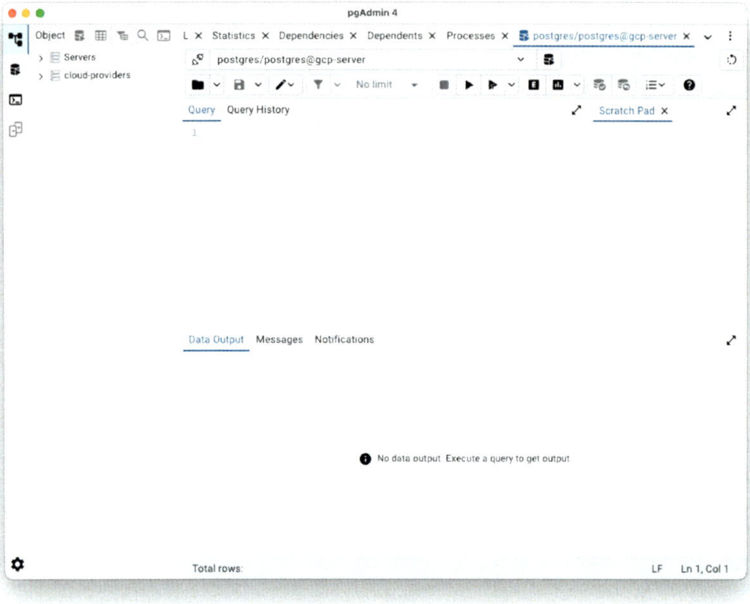

{width: 80%}

The pgAdmin interface is organized into several key areas:

- **Object Browser Panel** (left side): Navigate through servers, databases, and database objects
- **Main Dashboard** (center): View server statistics, recent connections, and quick access tools
- **Properties Panel** (bottom): Display detailed information about selected objects
- **Query Tool**: Integrated SQL editor with syntax highlighting and execution capabilities

Initial Server Connection

After installation, you'll need to register your local PostgreSQL server with pgAdmin. This creates a connection profile that pgAdmin uses to communicate with your database.

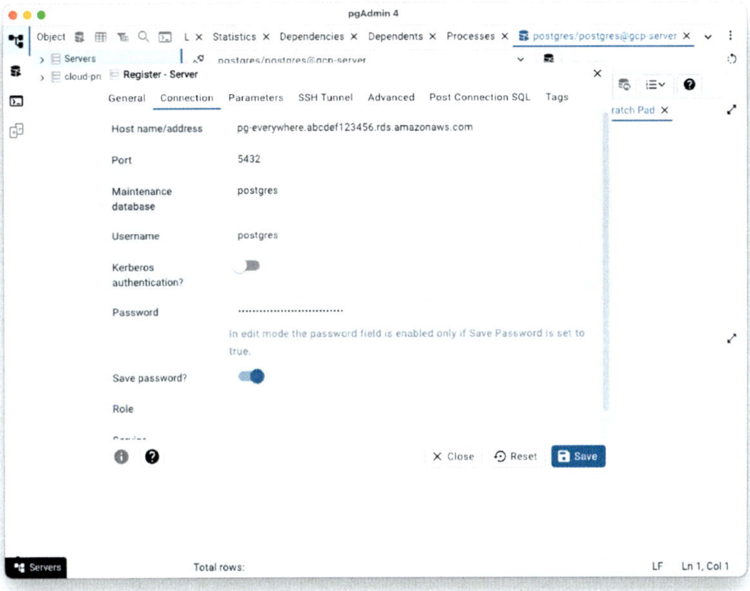

Server registration dialog showing connection details form with host, port, username fields

Connecting to Your Local Server 1. Right-click "Servers" in the browser panel 2. Select "Register" > "Server" 3. In the "General" tab, provide a meaningful name (e.g., "Local PostgreSQL") 4. In the "Connection" tab: - Host: `localhost` (for local connections) - Port: `5432` (default PostgreSQL port) - Username: `postgres` (or your created username) - Password: Your PostgreSQL password

Database Browser and Navigation

Once connected, the browser panel becomes your primary navigation tool, displaying a hierarchical view of all database objects.

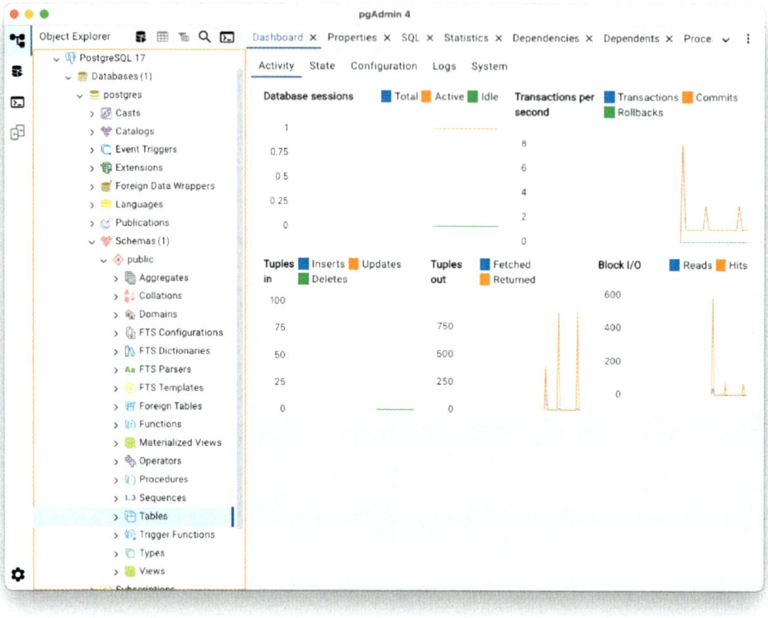

Expanded database browser showing databases, schemas, tables, and other objects in tree structure

Database Hierarchy Structure: - **Server Group**: Container for related servers - **Server**: Individual PostgreSQL instance - **Database**: Individual databases within the server - **Schemas**: Logical containers for database objects - **Tables**: Data storage objects - **Views**: Virtual tables based on queries - **Functions**: Stored procedures and functions - **Sequences**: Auto-incrementing number generators

Query Tool Features

pgAdmin's Query Tool is where much of your daily database work happens. It combines the power of psql with visual enhancements.

Key Query Tool Features: - **Syntax Highlighting**: Color-coded SQL for better readability - **Auto-completion**: Intelligent suggestions for tables, columns, and SQL keywords - **Query History**: Access previously executed queries - **Execution Plans**: Visual representation of query performance - **Data Export**: Export results to various formats (CSV, JSON, XML)

Query Tool Toolbar Options: - Execute query (F5) - Execute current statement - Stop execution - Clear query editor - Open/Save query files - Export data options

Database Object Management

pgAdmin excels at visual database object creation and modification, providing wizards and forms for complex operations.

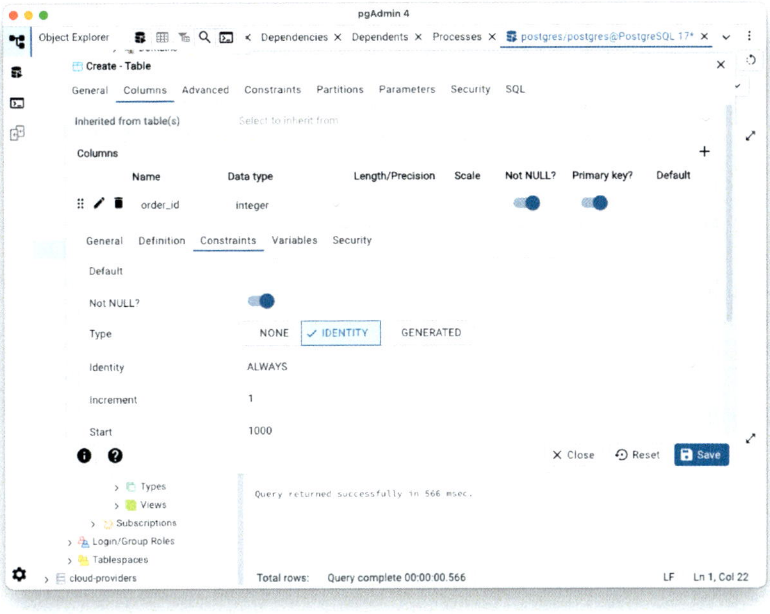

Table creation wizard showing column definitions, constraints, and other tabs

Table Creation Features: - Column definition with data types - Primary key and foreign key constraints - Index creation and management - Table permissions and ownership - Advanced options (inheritance, triggers)

Backup and Restore Operations

pgAdmin provides comprehensive backup and restore functionality through its graphical interface.

Backup Features: - Full database backups - Schema-only or data-only backups - Custom format for efficient storage - Scheduled backup options - Compression and encryption options

Restore Operations: - Point-and-click restore process - Selective object restoration - Role and privilege restoration - Progress monitoring and logging

Server Monitoring and Statistics

pgAdmin includes powerful monitoring tools to help you understand your database performance.

Monitoring Capabilities: - **System Statistics**: CPU, memory, and disk usage - **Database Activity**: Active connections and running queries - **Performance Metrics**: Transaction rates and lock information - **Storage Analysis**: Database and table size information

User and Role Management

Managing database security is streamlined through pgAdmin's user and role management interface.

Security Management Features: - User creation and modification - Role-based permissions - Group membership management - Password policies - Connection limits and restrictions

Query Performance Analysis

pgAdmin's EXPLAIN functionality provides visual query execution plans, making performance optimization more accessible.

Performance Analysis Tools: - **Visual EXPLAIN Plans**: Graphical representation of query execution - **Query Statistics**: Execution time and resource usage - **Performance Recommendations**: Suggestions for optimization - **Index Analysis**: Index usage and effectiveness

Practical Workflow Integration

Daily Development Tasks: - Use the Object Browser for quick schema exploration - Leverage the Query Tool for interactive SQL development
- Employ visual query building for complex joins - Utilize data export features for reporting and analysis

Administration Activities: - Monitor server performance through the dashboard - Manage backups and maintenance through scheduled tasks - Handle user management and security settings - Analyze query performance with visual EXPLAIN plans

Best Practices for pgAdmin Usage

Efficient Navigation: - Bookmark frequently used databases - Use the search functionality in the Object Browser - Customize the interface layout for your workflow - Take advantage of keyboard shortcuts for common operations

Query Development: - Save frequently used queries for reuse - Use the query history feature to track your work - Take advantage of auto-completion to reduce typing errors - Regularly review query execution plans for performance optimization

pgAdmin 4 bridges the gap between command-line efficiency and visual convenience, making it an essential tool in your PostgreSQL toolkit. Whether you're performing routine maintenance, developing complex queries, or analyzing performance, pgAdmin provides the visual tools to work more effectively with your databases.

Alternative GUI Tools

DBeaver

DBeaver (dbeaver.io/download) is a powerful, open-source universal database tool that supports multiple database platforms. It provides a rich set of features including an advanced SQL editor with syntax highlighting, visual query builder, and data visualization capabilities. For PostgreSQL specifically, DBeaver offers comprehensive support for native database features, including custom data types, stored procedures, and advanced PostgreSQL-specific SQL syntax.

DBeaver offers: - Multi-platform support - Advanced data editing - Visual query builder - ERD visualization

TablePlus

TablePlus (tableplus.com/download) takes a different approach, focusing on providing a native, lightweight interface with exceptional performance. It's designed with a clean, modern UI that reduces complexity while maintaining powerful features. While it's not free like DBeaver, TablePlus offers a streamlined workflow with features like code snippets, secure credential management, and multi-tab query editing. Its native implementation means faster response times and lower resource usage compared to Electron-based alternatives.

TablePlus provides: - Native performance - Clean interface - Multiple database support - Code snippet management

How does PostgreSQL official tools compare with others?

Each tool has its strengths. Many developers use a combination - pgAdmin for administration tasks, DBeaver for development, and psql for quick operations.

pgAdmin Core Strengths - Official PostgreSQL GUI tool - Comprehensive database administration features - Built-in ERD designer - Visual query explain plans - Server monitoring and statistics

DBeaver Distinctive Features - Advanced data editing capabilities - Visual query builder - Cloud database explorer - Grid colorization and data highlighting - Spatial data visualization - AI assistant for query writing - Schema and data comparison tools - Built-in task scheduler

TablePlus Key Advantages - Native performance - Clean, minimalist interface - Code snippet management - Multi-database platform support

Feature Comparison Table

Feature	pgAdmin 4	DBeaver	TablePlus
Query Tool	Yes	Yes	Yes
Visual Explain	Yes	Yes	Limited
ERD Designer	Yes	Yes	Limited
Data Import/Export	Yes	Yes	Yes
Cloud Integration	Limited	Advanced	Basic
Performance	Moderate	Good	Excellent

Usage Recommendations

Development Work - DBeaver excels for developers with its comprehensive feature set and data visualization capabilities - TablePlus is ideal for those preferring a lightweight, fast interface

Database Administration - pgAdmin 4 remains the go-to choice for PostgreSQL-specific administration tasks - DBeaver offers robust administrative features with additional monitoring capabilities

The choice between these tools often comes down to your specific needs and workflow preferences. Many professionals use a combination of tools, leveraging the strengths of each for different tasks.

Practical Workflow Integration

Let's discuss how to integrate these tools effectively in your daily workflow:

For development: - Use psql for quick schema checks and data verification - If you are not a console-guy or console-girl, leverage pgAdmin specially for complex query development - Employ GUI tools for data visualization and exploration

For administration: - Monitor performance through pgAdmin's dashboards - Execute maintenance tasks via psql scripts - Manage backups through GUI interfaces

Best Practices

As we conclude, let's cover some essential best practices:

- Learn basic psql commands even if you prefer GUI tools
- Use pgAdmin's query tool for complex query development
- Keep psql scripts for routine maintenance tasks
- Regularly update your tools to access new features

Remember, the best tool is the one that makes you most productive. Don't feel pressured to use only one - different situations call for different tools.

In your upcoming database work, you'll likely find yourself switching between these interfaces based on the task at hand. This flexibility is one of PostgreSQL's strengths - it provides powerful tools for every use case, from rapid command-line operations to comprehensive visual management.

About the Author

Jose Berardo is a senior software engineer, educator, and technology leader with over 25 years of hands-on experience architecting, building, and scaling systems worldwide. Jose's journey began with code in the late '90s and has taken him from small startups to global finance and SaaS providers. He's worked with every major database system—and discovered his passion for PostgreSQL back in 2001.

Jose's PostgreSQL expertise spans local dev envs, prod cloud deployments, distributed architectures, migrations, and performance tuning. He's delivered in-person and online training to hundreds of engineers, teams, and IT professionals, always praised for making complex topics both accessible and practical.

When he's not optimizing databases or writing for the PostgreSQL Mastery community, Jose shares tech tips and inspiration on Instagram @joseberardo and connects with database professionals on LinkedIn.

Ready to join thousands accelerating their database skills?
Enroll at postgresql-mastery.com!

Ready for More? Continue Your PostgreSQL Journey!

Congratulations on reaching the end of "Postgres Everywhere." But the adventure doesn't stop here. **Level-up with our hands-on course at postgresql-mastery.com.**

- Master advanced SQL, performance tuning, and DevOps workflows
- Access exclusive video tutorials and real-world deployments
- Join a vibrant community of PostgreSQL pros and beginners
- Get personalized feedback from Jose Berardo and our instructors

Let's Connect!

Me, having fun somewhere

- **Instagram:** @joseberardo
- **LinkedIn:** linkedin.com/in/joseberardo

Your database mastery is just beginning.
See you in the course!

Made in United States
Orlando, FL
05 November 2026

90200302R10105